D0952033

BRAVE

CULT MEMBER. RUNAWAY. CAPTIVE. STARLET.
VICTIM. SEX SYMBOL. JUSTICE SEEKER.

BRAVE

ROSE McGOWAN

HarperOne
An Imprint of HarperCollins*Publishers*

Some names have been changed in this book to protect people's privacy.

 For information, address HarperCollins Publishers, 195 Broadway, New York, NY 10007.

HarperCollins books may be purchased for educational, business, or sales promotional use. For information, please email the Special Markets Department at SPsales@harpercollins.com.

FIRST EDITION

Designed by Yvonne Chan

Library of Congress Cataloging-in-Publication Data has been applied for.

ISBN 978-0-06-265598-1
ISBN 978-0-06-285178-9 (BN)
ISBN 978-0-06-285179-6 (BAM)

18 19 20 21 22 LSC 10 9 8 7 6 5 4 3 2 1

CONTENTS

Dedicated to all of us survivors

AUTHOR'S NOTE

My life has always been one of extremes. *BRAVE*, the book, has proven to be no different. While writing this book, I endured being hacked, stalked, spied on, had parts of this manuscript stolen. My life was infiltrated by Israeli spies and harassing lawyers, some of the most formidable on earth. These evil people hounded me at every turn while I went about resurrecting the ghosts that have made up my time on earth. I can only say it was extraordinarily stressful, an incredible high-wire act that required great strategy. There was never any other choice. Justice would be served.

And it was. I am immeasurably proud of having a hand in this cataclysmic global reckoning and the felling of monsters. I truly believe that a win for one of us is a win for all of us.

A few years ago, I realized society needed to be primed to

hear The Story, so I set about taking my voice of dissent public. I decided to openly fight the machine, the manufacturers of myth, the gaslighters themselves, the sacred men of Hollywood. For far too long they'd been on top and able to get away with criminal behavior. I wanted to make it impossible to look away. And then the US election happened, making sexism far harder to deny; it paved the way for obvious truth to be revealed to those who'd for so long turned a blind eye.

In early 2017, I'd been working on *BRAVE* for a few months when I made contact with two investigative reporters. It was time. The story took many twists and turns as it all unfolded, and I'm proud to have had a hand in starting the worldwide conversation.

Since I and so many brave survivors have come forward, titans of every industry have toppled. We survivors have gained our power. We survivors are using our voices in record numbers. We cannot let up, and as hard as it is, we must continue to get even louder, to push even harder. We all count. We all matter.

Here's to freedom, yours and mine.

Now go breathe fire.

RM

PREFACE

"Did you break up with someone?"

At first the question made me angry. I thought it sexist, stereotypical, disheartening. There was no death of a relationship that made me so in need of freedom that I'd alter myself. The more the breakup question was asked, the more it made me think about my motives. I realized I *had* broken up with someone. I broke up with you. The collective you, the societal you. I broke up with the Hollywood ideal, the one that I had a part in playing. The ideal version of "woman" that is sold to you by every actress in every hair commercial telling you, "This is the secret to being beguiling, the secret to getting a man to want you." Long, glossy Kardashian-esque hair that says, "Fuck me, big boy." As if that's all we are and all we can be. Hair. Hair is what I broke up with. And it was a breakup that was years in the making; it took a lot to wake me from my brainwashed slumber. My long hair had always made me uncomfortable. It made men look at me while the real me disappeared. I would use it to cover my face, to check out, to sleep. And sleep I did. The real Rose slept while the fake Rose lived a bizarre alternate life playing the part of someone who played parts.

Most of my life I had short hair. I preferred it that way. The classic film stars and punk women I most admired had short hair. I liked very much being an individual. I liked looking neither female nor male, but hovering somewhere in between. The two periods of time when I had long hair were the hardest in my life, the times I was most lost from myself—my teen years when I suffered from a raging eating disorder and later when I suffered from a mental disorder called Hollywood. The Hollywood disorder lasted a much longer time, but both had to do with being absent from self. Both times were driven by society's number one propaganda machine—Hollywood. I was told I had to have long hair, otherwise the men doing the hiring in Hollywood wouldn't want to fuck me, and if they didn't want to fuck me, they wouldn't hire me. I was told this by my female agent, which is tragic on many levels. So, so evil and so, so sad. Evil because I took the information from an older woman who was the mouthpiece for what Hollywood wants. Sad because she was right. This message gets filtered down to all women and girls, telling us to have long hair so we too can be sexy, but

I got the direct message, like a hotline phone call straight from what "the man" wants.

Well, fuck Hollywood. Fuck the messaging. Fuck the propaganda. Fuck the stereotypes.

If you're a Jennifer Lawrence, America's sweetheart type, you have simple blond hair. If you're the vixen, it is long, dark, and big. Those are the rules, do not deviate. My long hair was beautiful, like beauty pageant contestant hair. My hairdressers were gay males and I was their Barbie come to life; at least that's what they told me. I didn't think I looked like Barbie. I thought I looked more like a blow-up sex doll, the kind with the hole for the mouth. I had been turned into the ultimate fantasy fuck toy by the Hollywood machine. All the men and women hired to make me look like said fantasy fuck toy did a good job, but I was dying on the inside and embarrassed by what I looked like on the outside. But I didn't know how to change what was wrong when there were so many levels of wrong in my life.

I meet so many women and girls who tell me their hair is a security blanket and what they hide behind. I find this not only relatable, but heartbreaking. Of course you should have long hair if YOU feel like having long hair, but examine your motives. What part does society play in telling you how you should look? What part does media play in showing you what you should be? And if you are hiding behind your hair, why do you want to live a life in hiding and what are you hiding from?

When I shaved my head, it *was* a battle cry, but more than that it gave me an answer to the question I so hated.

Did I break up with someone?

Yes, I broke up with the world.

You can, too.

My name is Rose McGowan and I am BRAVE.

BRAVE

INTRODUCTION

There once was a famous actress named Frances Farmer. She hated everything about her artificial life. She wanted to be free. Frances tried to escape fame and the toxicity of Hollywood's male-dominated world, but the studio had her captured. They took Frances to a mental institution. They locked her up. There was nothing wrong with her mind, she just didn't want to be famous. She screamed, begging for her life. Instead they took it. They laid her down, restrained her, and shocked her mind with electricity. Shock. Shock. Shock. Over and over. The male powers that be in Hollywood wanted Frances to be a submissive good little girl, and remain so. What they left of her was an empty shell, a husk of a woman. Frances was never Frances again. And all because she didn't want to be sold as entertainment.

Very few sex symbols escape Hollywood with their minds intact, if they manage to stay alive at all. The streets of Hollywood are paved over the bodies of the vulnerable, the fucked with, the lied to, and the hurt. I know, I was almost one of them. You may think that what happens in Hollywood doesn't affect you. You're wrong. My darlings, who do you think is curating your reality? Who is showing you who and what you want to be?

I want to have a frank conversation about an inner sickness that I see few, if any, addressing: how and why Hollywood creates a fucked-up mirror for you to look in. How you are seeing yourself through your own eyes, but perhaps not your own mind. Hollywood affects your life in ways you may not even be aware of.

In my past of being sold as a product, I have been a part of massaging your brain. I wiggled into your mind professionally. I was the cigarette the advertisers told you you needed. I've also been on the other side of the looking glass. Watching you. Studying you. Impersonating you. All of us in Hollywood, media, and advertising do. And you know what? We are really good at it. We have had it drilled into us how best to be marketed to you. How best to be sold to you. How to implant what "we" want into your brain, into your thoughts, into your wallet. And it works. You're sold a fake reality all for the rock-bottom price of $14.

The men who thought they owned me think that they own you. They are the latest in a long line of myth peddlers, from the men behind the Bible to these modern-day "content creators."

They're mostly self-aggrandizing, egomaniacal abusers of power. And they've never been more dangerous. Few in Hollywood, and no actress that I can recall, has gone rogue. Hollywood operates like the Mafia when it comes to protecting its own. Especially if your "own" is a rich white male. Yes, I said it. But here's the thing, it's true. I didn't make it so, it just is. In other news, the sky is blue.

By telling some of my story, I aim to shine a light. For those who think Hollywood is a silly joke . . . it's not. It's a deadly serious business and one that keeps its winnings. You may think it's as simple as forking over hard-earned cash for a night out at the movies or paying a cable bill to be entertained. I'm here to tell you the price you are paying is much higher than you know. You are paying with your mind, your behavior, and your patterns. Things that should have no price tag. In our as-seen-on-TV society, the simple fact is that what you have watched and consumed, from birth, has formed you and continues to form you. Even those who've opted out of its false reality have to stay vigilant to remain free from the lies and from the messages that do far more harm than they should. Because they are insidious, and they are everywhere.

My life, as you will read, has taken me from one dangerous cult to another, one of the biggest cults of all: Hollywood. I say biggest because short of a nuclear bomb, Hollywood has the farthest reach. *BRAVE* is the story of how I fought my way out of these cults and reclaimed my life. I want to help you do the same.

You can say "no more."

You can say "yes" to a freer you.

You can be free of the trap that's been set for you. And believe me, it has been set.

I am writing this book because I want to have a real conversation with the public and most especially you. I am honored that my words will enter your consciousness and conscience, that my thoughts will rest in your mind. I take that responsibility seriously.

Call what I'm doing a public service and you'd be correct. It is.

Hollywood is a dirty town up to some dirty tricks.

This is not a tell-all.

This is a tell-it-how-it-is.

PART ONE

CHILD OF GOD

Here's the thing about cults: I see them everywhere.

If you're deep into the Kardashians, you're in a cult. If you watch your favorite TV show and go online and you're in chat rooms with everybody else who's obsessed with that show and you're breaking it down episode by episode, you're in a cult. If you're bingeing, scrolling, absorbing from one news source more than any other, especially if it happens to be fair and balanced, you are in a cult. You're living your life through other people. If you blindly vote for so-and-so, you're in a cult. If you're deep into your country's propaganda machine, you're in a cult. Look around you and see where the cults are, because they are everywhere. Anywhere there is group thought and group mentality: you're in a cult, you're in a cult, you're in a cult.

The first step to deprogramming yourself from a cult is realizing you *are* in a cult. I would know, I escaped from two of the most iconic cults of all time.

For those who knew me as an actress, I must inform you that I was never that person. I was playing the part of someone who played parts. I was trapped by rigid societal ideals and gender expectations placed on me by people who shouldn't have been allowed near me (or you). I got such a deeeeeeeeep mind fucking. I rejected brainwashing early on in life, but later, Hollywood's Cult of Thought actually got me.

My life altered irrevocably the day I turned into a pixel, beamed up to an orbiting satellite and beamed back down, blasted across living rooms, bedrooms, lives. My job was to take you away from your struggles for a while, to make you feel empathy, to make you feel at all. I took my job seriously. But like in most cults, because I was a woman, I was considered to be an owned object. I was sold for the pleasure of the public. Deeply programmed men (and women) made money selling my breasts, my skin, my hair, my emotions, my health, my being. I was not taken seriously, nor was I respected. Not by most of society, and certainly not by the Hollywood cult with its massively industrialized Madonna/Whore complex.

Imagine if your value to the company you work for was measured by how much semen you could extract from anonymous masses of men. 'Cause you know, if strange men masturbate to your movies, you must be of some value. Sounds like a sex worker, right? You're not too far off.

Imagine that every word to come out of your mouth for nearly seventeen years, day after day, month after month, angle after angle, take after take, was something an all too narrow-minded male wrote for you to say. It's meta and it's deeply abnormal.

It took me a long time to figure out that I was in another cult, because I was too busy being other people, not myself. By telling the story of my life, I am reclaiming it.

But let's start at the beginning, shall we?

In a stone barn, in the tiny Italian countryside town of Certaldo, delivered by a blind midwife, as the story goes, I came into the world. There's an American saying: "Shut that door! Were you born in a barn?!" I guess I never have to shut doors if I don't want to. I have that prerogative. I suppose sometimes you're just earmarked for weirdness from birth, and I think I'm one of those.

The barn was on the property of the duke of Zoagli, known as Duke Emanuele, who, upon joining the Children of God, donated his estate and land to Children of God. His sister Rosa Arianna lived on the property, but loathed all the Children of God members living there. My parents named me after her, Rosa Arianna, I think to make her like them. Didn't work.

It was incredibly beautiful there in the rolling hills outside Florence, the dark green cypresses and silvery-green olive trees, vineyards, and orchards, those enormous old terra-cotta jars holding red geranium flowers. I suppose if you have to be in a cult, it was as good a place as any.

Nah, it was better, and even at a young age, I saw the beauty and knew it was wildly extraordinary. I connected to its nature as an escape from what I was born into. As a result, I've always been drawn to shapes, colors, and light patterns, and the Italian countryside has haunted me my whole life, in a good way.

From my earliest memories I recall hearing a lot about a terrifying old man named "Moses" David Berg, our fearless leader in the Children of God. He would send his directives out in cartoon pamphlets called "Mo Letters." Whatever Moses David wrote, that's what was done. Each time there was a new letter it would be as if the ruler of the universe had spoken. (Kind of like the head of a studio in Hollywood.) And I guess as the self-appointed prophet he was, Moses David turned out to be the King of Creeps. But the others didn't know that yet. Some would never know.

I remember a lot of hairy legs, men's and women's, like in the cartoons where you only see the adults' legs because that's your perspective as a child. I remember a lot of singing, praying, clapping, and snapping. Yes, snapping. I was told I had to sit on the floor all day and learn how to snap my fingers, otherwise God wouldn't teach me to drive when I was sixteen. I didn't understand anything about sixteen and driving, but even then I could tell finger snapping as the key to doing anything was patently absurd.

One night, a ghostly looking woman in a white robe came into the room I was in. She was like a shadow holding a candle—there was no electricity. It was storming outside and I remember the wooden shutter slapping against the old glass window. I had been worried the window was going to break, but I was now distracted by the woman in white who sat by my feet. The wind was whistling through cracks in the stone and I was having trouble hearing her. The wind stopped and she looked straight into me and said, "Have you let God into your heart?"

I sat up, looked at her, considered carefully, and shook my head no.

The woman pinches my foot and twists my skin. I am not going to cry out because I know that's what she wants. For this refusal there was punishment. Corporal punishment, slaps and spankings, because "spare the rod, spoil the child." She twists harder. I bite the inside of my lip so I don't cry. I stare back, silently defiant.

The woman says it again, this time in German, "*Hast du Gott in dein Herz gelassen?*"

I think about it and say, "No. Not today. Try tomorrow."

She slaps me across the face. Hard.

Even at that tender age, I reasoned that if I invited him into my heart, it would be their God I was letting inside. It would no longer be my God, whom I was very protective of. And their God was cruel. What they were preaching made no sense to me, their actions not squaring with their words. That was not a reality I wanted to exist in.

Later my younger sister Daisy urged me to just say yes, that it would go easier for me, but I kept taking the punishment instead. I was, as my name foretold, quite thorny, whereas my sister was a little golden-haired, sweet child. I would stare at her and wonder how she got that way and how she couldn't see what was going on. It was a strange sensation growing up behind these walls and being told I did not belong to the outside world, but I also knew I didn't belong to the world within.

When that woman or another woman or another man, all strangers, returned the next night and the night after, I always had the same response: "No, no, I have not let God into my heart."

Slap.

One night I could hear the woman's German whispers and her feet doing a quiet kind of stomp on the floor. I knew I was going to get hurt again.

"No."

Slap.

When she was gone, I saw that she left her Bible on my sleep mat—all the kids slept on flimsy orange or blue plastic mats. I hid her Bible behind a cabinet. Each day I'd tear out a new page, put a small piece in my mouth, work it around, add more, and spit it out, turning it into little mush blobs. Then I would take the Bible blobs and form them into tiny animals. I hid them behind the cabinet and would visit them now and then when I could steal a moment. They were my toys, one part saliva, one part Jesus.

I figured if I literally ingested their God, maybe I could answer, *Yes, I have let him in*. Maybe they'd stop punishing me.

The smacks, the pushes, enforced the message that you were not allowed to be imperfect. When I was about four, I had a wart on my thumb. I was toddling down this long hallway when one of the doors opened. I remember the shaft of light and all the dust motes dancing. A man with shaggy blond hair picked me up, looked at my hand, and said, "Perfection in all things." He held up a razor blade and sliced my hand with one swipe, winking at me as he sat me back down. "Perfection in all things," he said again before shutting the door and leaving me in the hallway. I didn't cry, I was too stunned. Blood ran over my hand and I made a dripping mess of the hallway. The blood coursed over my fingers, the red strangely pretty. Like my hand, I was numb. I knew not to react because, one, that was something they wanted from me, and, two, I thought maybe there was something to this perfection thing. I walked on.

The hallway assault is what started a narrative that fucked with my head for years, that of perfection as self-protection. I told myself if I were just perfect enough, I'd be okay. If I were just perfect enough, I'd be left alone and no one would want to hurt me.

From then on, I willed myself to be as perfect as possible because I didn't know what would happen to me if I wasn't. I was terrified of having an aberration in any way. I was sure that having any kind of flaw would spell doom. But first I had to figure out what all my flaws were. And so began a habit of being extremely hard on myself, seeing myself from the outside in. I started to look at my hands and feet daily to make sure I didn't have any bumps growing. There were no mirrors that I can remember in the cult. When I would later arrive in a culture that was so externally focused—America, and then Hollywood—this caused a tear in the fabric of my being.

The funny thing was that in almost direct opposition to the message the cult sent us about perfection, my father was preaching to me and my siblings that we were not, under any circumstances, to develop an ego. Our focus was to be on our internal development, the development of our souls and our intellects. I suppose we were supposed to be perfect physically, but remain humble in the face of our perfection? I was never really sure. All I knew was that I was not supposed to think good thoughts about myself. That God would punish me for thinking that I was awesome.

Never once growing up was I told that I was intelligent, smart, or beautiful. I don't know what that feels like. I was never told I could do anything I wanted if I set my mind to it. I was told I was worth nothing in the eyes of God. I was told I was going to be a whore. I was told I was dirty. And the thing is, I knew they were wrong, but the words still stung.

From an early age, I remember being furious that nobody would listen to me just because I was a child. It was so unfair. I hated being little and powerless. I would look at the people in Children of God and think, *But all these things you're all talking about, I could solve them in two easy steps if you adults would just listen to what I am saying*, but nobody would listen to me. Because I was a girl. That set a real pattern for my life. I was a born dissenter—not for the sake of being contrary, but because if you could see things for what they were, identify the source of a problem and the solution, why wouldn't you want to fix it? But nobody would listen to me. They just sat me at the little kids' table. Not unlike later in Hollywood. Just a girl, after all.

My only friends during my time in Children of God were my older brother, Nat; my pet lamb, Agnello; and an old gray-haired farmer named Stinky Fernando. Stinky Fernando was deeply suspicious of bathing. You could almost chew his smell, it was so thick. I had to breathe through my mouth whenever he was around. One day I heard Stinky Fernando screaming. My father and some of the other members took him by his arms and ankles and threw him in a river. Much to Stinky Fernando's surprise, his skin did not melt off.

Stinky Fernando took Nat and me into an old barn and showed us faded *Playboy* magazines while feeding us stale Kit Kats. A real treat. I wondered about the women in the magazines. They didn't have hairy legs. It was confusing. I loved the rancid Kit Kats, though. I loved candy way more than I loved their God.

I bottle-fed my friend, the little lamb Agnello, and helped take care of her. My first pet. One night at the long dinner table I took a bite of food, and a thin woman with a mean face and center-parted hair started to laugh. Others joined in, and soon everyone was laughing. I didn't understand what was funny until they told me it was Agnello being served. And so I realized my pet was being fed to me for dinner. I sat stunned while everyone at the long table laughed. I pushed my tears down and felt a coldness wall off my heart toward these people, something crystallizing into a stone of pure hatred as I looked at their monster faces. They had a particularly cruel streak, and they liked to destabilize the younger members. These were lovers of Christ, right? To this day, I've never eaten lamb again.

I started to become angry. Angry at the injustices that were adding up. Angry at the rules that seemed, and were, so arbitrary. I decided the best course of action was to light it up. And so, one day my older brother decided to light a stable on fire. He was mad, too. I for sure wanted to be there for that, so I ran after him to help. We were in the barn when my brother pulled out a book of matches. He started lighting them and flicking them at the hay on the stone floor. *Whoosh.* The fire leaped up the side of the walls and onto the ceiling. The roof was thatched hay and started popping above us. I tried stamping out the flaming pieces with my feet, but I was too little and it was too late. I stamped and stamped, but I couldn't put them out. If I had known how to say *fuck*, I am sure I would have. The roof crackled more and it was getting very hot. I knew we were in big, big trouble if we went outside and were caught by the adults. But everything was on fire.

We chose to run.

This is the part where I'm supposed to tell you some hideous story of punishment for lighting it up, but I really can't remember. I do remember the terror of being found out. It made me feel like my skin was about to fall off with fear. The movie scene of this would be:

A sturdy blond boy and an elfin girl are hiding from their father. Suddenly four hands grab them by the shirt collars, dragging them off. Turning down a path in a maze, the children are paraded past a gauntlet of leering cult members. The members drag the children to the Judge of All. The Judge of All is on a throne made of soft wood. There are young nude women, heavy breasted and round bottomed, on their knees, gazing up adoringly and reverentially at the dynamically dangerous leader. The leader tilts his head back, eyes shut. He's being worshipped. He's in heaven on earth. The women work oils and lotions into the leader's skin, their hands using a feathering touch as they go, chanting with intention. The leader, the Judge of All, opens his eyes and points at the boy and girl. The shaming begins.

Sounds like a Hollywood film, right? Maybe it's not too far off. In fact, my life as a performer began there in the cult. We were made to go out in groups to sing at local orphanages and hospitals, or on the streets, to perform. Singing Jesus songs on the streets of Rome with a hat in front of me, street busking. After the coins would stack up in the hat, a hand would come over my shoulder to take all the coins I'd earned. They let me carry the empty hat. Gee, thanks. It was my work that was bringing the money in and I was pissed at the injustice of having to give it up. I'd see regular families with the kids walking around with gelatos and candy and I'd wonder about their lives at home. Did they have a bed? We had plastic mats and I got cold at night. The girls wore pretty dresses; I had faded brown overalls and Jesus sandals. My hands and feet would get dirty and I'd try to hide them when other, cleaner children looked at me. For hours we would stand and sing those damned songs, under hot sun, in the rain, it didn't matter. I was five or so. My little legs would get so sore from standing, but I knew I couldn't sit or there'd be trouble.

We had to return with money or else there would be sanctions and punishments against our family. I could feel the stress of the adult members as the "Systemites" (that's what they called people outside the cult) turned away and ignored us and the pamphlets we were selling. Little incoming money equals not much food. Not surprisingly, there was often hunger. Our food was rationed. If we returned with not enough money, the rationed food was given to another family as punishment. If potential new members or press were coming to visit, they'd put us kids on a white rice, milk, and sugar diet to fatten us up. We'd stuff ourselves with it until we gagged, but I loved it because at least there was something to keep me full. Plus, sugar, which I loved.

Sometimes the local press would be invited to come and cover our good deeds: "To see what great work we're doing in the Children of God community, join us." See, we're not a bunch of freaky hippies, what kind of freak could sing a Jesus song this well?

I was sent to entertain sick children in hospitals. I remember thinking: *Kid, if this is your last day on earth, I'm really sorry that we're forcing you to listen to little me singing about Jesus. I don't want to be here, either. I apologize.*

But even though it was awkward performing in hospitals—and this may sound weird—I always knew I was going to be famous, even before I understood what fame was. It was kind of a foregone conclusion. I don't know how to explain it.

At some point in my childhood I remember being taken to see a film. It had a great impact on me. I don't know what it was called. It was Italian. The lead actress had short raven hair and was a nurse. She wore a crisp white uniform and a little white hat. She was in a phone booth, crying and screaming at her married doctor lover, who was throwing her aside. She took the back of her hand and smeared her lipstick across her face. She ripped her shirt open, popping its buttons. Her chest exposed, she took lipstick out of her purse and drew all over her breasts like a wild woman. I was captivated. It was fabulous. I wanted her lipstick and her hair. I finally got to see some glamour in my young life and I knew it was for me. My feelings of being in the wrong life intensified.

At some point my father found a Brownie, a vintage camera, so the few photos that exist from my childhood look like they're super old and are largely black and white. I watched my father as he captured objects and people with the camera. Then I got to play with it myself. I learned to see things through a frame. Looking through that crappy lens, I felt as if I could see more and that everything I looked at told a story. Soon I was nearly always outside of myself, watching and filming and documenting everything that was going on, taking note of everything: smells, sounds, tastes, situations, people. Only now can I see that this was early disassociation to deal with trauma. Looking through a lens has been a coping mechanism I have employed throughout my life. It had a silver lining: my falling in love with photography and cameras. But more than that, it gave me a way of putting something between me and the world, and a different way of looking at it. Every detail as seen through a lens. Because it's not really happening if I'm once removed, right?

I also used books as an escape. Words were my solace and my saviors when I was small and have remained so to this day. Words, different lives, different centuries, that was how I survived.

Books also furthered my training for being an actor because I took on the persona of whichever character I was reading. It could be a serf, it could be a queen. I would mimic the posture, everything about that character, while I was reading his or her story. When I finished a book, I went into mourning for that character because it was a death. I took books very seriously. But not the Children of God books. I could not understand how anyone could believe them. Those Mo Letters were just so . . . well, stupid. It's so hard to understand how so many have fallen for it.

Meanwhile, the beliefs and practices of Children of God

started getting more and more dangerous. Moses David, our leader, made the young women members do this thing called "flirty fishing." He sent them out—and these were little more than girls, really—to seduce men at bars or cafés. The men would wake up in the cult. Moses David christened the girls "Hookers for Jesus." Hookers for Jesus? Fuck you, Moses David, you piece of shit. Fuck you for all the pain you caused. At the end of the day, it was all about male dominance, and using sex as a weapon for mind control. Beautiful women were major targets, not unlike what I would later see in Hollywood. And, like in Hollywood, there were women who helped Moses David do bad things to others.

The cult was a highly sexualized environment, run by men, to benefit men. My father loved it, I could tell. I remember standing in a corner, watching my father preach, as he sat on a thronelike rattan chair. Women—girls—were on their knees staring up at him with dreamy expressions. Women literally worshipped at his feet. I remember looking at the women on their knees. Then my father on his throne. *I'll never be like those women*, I thought. *Never.* It grossed me out. Looking back, it was the time of my father's life when he was at his most radiant. Abuse of power was inevitable, and he certainly abused his position.

One day my father said to my very young mother: "Saffron [my mother's name in the cult], I want to be married to this other woman as well." Well, hell. That must have sucked. There have been lots of times I have wanted to go back in time and kick my father's ass, this being one of them. My poor mother's own mom, Sharon, had just died tragically. My mother's dad was gone, too. She was alone in a cult in another country with a bunch of kids she was told to have and now this? It must have been crushing. She had no choice, and he took another wife. That's how my four youngest siblings—two full and two half—are so close in age.

Children of God next started advocating child-adult sex as a way to "live the law of love," which is just beyond disgusting and criminal. I saw an eleven-year-old girl being forced to sit next to a naked man, with his floppy dick on his leg. They made her sit between his legs so he could "massage" her back. I saw her tears. Even then I knew none of it was "normal," whatever normal was. I don't think there really is such a thing as "normal," but I knew that this was something deeply wrong, something to be avoided at all costs.

I feel bad for that small child I was, who from age three or four already knew so much about surviving. I didn't know what it was like to feel safe. In its place, there was stress and, underneath it all, a deep undercurrent of fear running through the commune. From a very early age, I realized kids were very far down the list on things to care about, which is lame when you're the child on the bottom of that long list.

An unfortunate necessity in this environment was being able to immediately pick up on danger. I excelled at it. One of my survival skills was, upon entering a room, to locate a weapon. I would do an immediate scan of the area to see what

I could use to cause someone else the most damage and defend myself against attack. My quick mind and rapid thought processes have been my lifelong savior as much as my fight. I've always gone by the seat of my pants, and my intuition is damned good. It's too bad I didn't apply the same skills to Hollywood. It would have saved me a lot of heartache. It could have saved me from unspeakable trauma.

In any case, my outwit-and-outlast mentality served me well as a child. Thankfully, I was just young enough to escape getting molested, or maybe my penchant for always having very short hair and wearing my brother's hand-me-downs helped save me. They thought I was a boy most of the time. Although the boys certainly got nailed, too. Fuck, maybe I was just too much of a troublemaker.

It would only have been a matter of time, but luckily for us, my father drew the line at pedophilia, and he made secret plans to leave. We couldn't just announce we were leaving and walk away, though. When the cult got wind of certain members wanting to leave, one of their children might disappear, or some family would get severe punishment meted out to them, as a way of teaching the others.

And so one night, my father told us there was a man named Bepo and he was after us with a hammer. There was a car waiting for us, we got shoved in, and go go go.

First we fled to a place called Munano, a small town in the Tuscany region of Italy. We lived in a centuries-old stone house where we boiled water and bathed in a round rusted metal tub.

We were scraggly kids wearing hippie hand-me-downs. I was used to having many kids around me, so it was strange to share a room with only four other children, to be suddenly with so few people, even if they were my actual family instead of "The Family."

My father had left the Children of God physically, if not mentally, taking his other wife, Esther. As for my mother, all I know is that she was left behind. There were so many women in the cult that I didn't have a firm grasp on my mother as an individual. It was just one more level of destabilization in what would be a pattern for me in my life.

Now that we were in this small medieval town, I was sent to my first public school. It was very confusing. In the cult, we had worn whatever fit from the pile of donated and hand-me-down clothes, and I mostly wore my brother's clothes. Now I was assigned to wear a pink button-down smock. I preferred the blue smock and asked why I couldn't wear it instead. I asked the teacher about this logic, and she told me because I was a girl I had to wear pink. Only the boys wore blue. I thought that was some of the dumbest shit I had ever heard. I was furious that I was now different from my brother for an arbitrary reason. I didn't understand why I now had to wear pink. I still don't.

We had a neighbor lady named Antonella who knitted me yellow wool underpants, which are just as uncomfortable as they sound. They were bulky and lumpy and needed to be tied so that they stayed up. I was and am wildly allergic to wool. They itched so much on my way to my first day of Systemite school that I ditched them and left them behind a bush. I walked out onto the play area on my first school break. It was stressful for me because I didn't know what kids did in regular school. I didn't know what the bells meant or what the rules were. That day, the girls on the playground were doing the Italian version of Ring Around the Rosie: *Amore, Tesoro, Salsiccia, Pomodoro!* After saying "pomodoro," the little girls would fall on their backs and kick their feet up in the air. I saw a girl approaching to invite me to play. I thought about my lack of underwear and stiffened, pulling my stupid pink smock down farther. The girl came closer and asked me to join the others. I went mute. I shook my head vigorously side to side, standing rigidly against the wall of the schoolyard. I pulled my smock down farther in case she had any idea to drag me out to play. The girl looked at me like I was crazy. I was still mute. It was my first interaction with a noncult child and I didn't know how to talk to her. So I just stayed mute.

I was immediately shunned by her and the other girls and labeled a snob. Sigh. It set the tone for the rest of my patchy scholastic career and, really, my later life. They knew I was different. I knew I was different. Not better than, not worse than, just different. It wasn't a feeling I had, it was just a fact. I didn't integrate well. I didn't relate to children at all. Theirs was a language I didn't, and couldn't, speak. They had concerns in life to which I couldn't relate; my problems were about surviving. When you have really big, dark things happen to you, it takes a lot more to care about things. It felt like I was about eight thousand years old in a small person's body, essentially an alien among those who understood traditional societal constructs.

On the way to and from the village school, I buried insects. I had a knack for seeing insects in peril. I would arrive at school puffy eyed from shedding tears at my self-made bug funerals. I'm sure I was unnerving as a child because of my intensity. I know I was because I basically was the same as I am now, and I tend to unnerve people to this day. I saw past everything, all the spiderwebs that people often hide behind so they can tell themselves a story about themselves. It made my father in particular very uncomfortable. Like I said, he may have left the Children of God, but the need to be a demigod preacher never left him, and suddenly he was a cult leader without a flock. He expected the women and children around him to worship him, and I never did. To have somebody around who's staring at you and puncturing through the falsehoods you've established to live your life must have been unsettling.

Thus, in an ironic twist after all he'd put me through, my father lectured me on how essential it was for me to be *more* childlike. Couldn't he see that he was the reason I couldn't be? When the onus of survival is put upon a child, surely that hinders her ability to go play with a doll like other good little girls. I could lay that squarely at my father's feet.

Going back through my life, I can see why my father lived his life differently; it was a reaction to how he was raised. Well, that and a uniquely special kind of mind. His father, my grandfather James Robert McGowan, was an American patriot, a navy man through and through. He was complex, alcoholic, and tough. Grandpa Jim had five children under the age of eight when he went to fight in Korea. His wife, Nora, a mentally fragile beauty from Quebec, Canada, was left to raise the kids alone while he was at war. When Grandma Nora found out that Grandpa Jim had a Korean lady on the side, her occasional bouts of mental illness became intractable. At least that's what my dad told me. The navy committed her to their psych ward, and involuntary electroshock therapy followed. It was early days for that kind of therapy: Grandma Nora was a guinea pig.

During this period, chaos reigned in my father's childhood home. My dad and his siblings would alternate wearing the one pair of shoes they had. Only the child wearing shoes could get to school and have a hot meal. It breaks my heart to think of them taking little forged notes from a "parent," written in a child's scrawl, asking the grocer to please give them bread that would be repaid later. A gang of men broke in one night when they were alone and ransacked the house. The children hid in the refrigerator.

I would like to say I can't imagine their terror, but it wouldn't be true. I can relate to the instability, hunger, raging mental illness and its fallout. These are all old friends of mine, and my family's.

I would later come to understand that my father was most likely manic-depressive. The manic side was the magical side, the bright, funny, wild side cackling at the wheel of our car as we skidded around the mountain roads. The depressive side was the monstrous, violent side. The more overwhelming his life got, the more the dark side won out. He had fallen into doing heroin at one point in the early '70s in Venice, California. I think that's where he met the people from Children of God, and how he got clean, and then his new drug became Jesus. He was like a rock star and Jesus became the instrument he played.

Had he gotten help earlier in life, it would no doubt have saved my relationship with him, and his relationship with my brothers and sisters, his relationship with art, his relationship with the world, with women, probably with everything. As for my mother, with her porcelain skin, long, reddish-blond hair, and blue eyes, she was a magnet for the wrong kind of boy. The wrong kind of boys turned into the wrong kind of men. She ran away by the age of fifteen. At eighteen she met my father, Daniel. By nineteen she was pregnant and in a cult.

While my mother was pregnant with me, her mother, Sharon, climbed the Three Sisters Mountain in Oregon and tragically slipped, plummeting to her death. She was thirty-seven. I was told that's why I'd always be sad, because my mother was sad during my pregnancy. For years I thought my intense internal sadness was due to this, but later I realized it had more to do with brain chemistry.

Sharon, with her beautiful red hair and green eyes, had also married young, a bad match. I've blocked out my grandfather's name, her husband. I suppose I could find out, but I frankly don't care to. The mores of the time being what they were, a damaging blanket of silence covered all intransigence. The dying gasps of the Kennedy era and the pervasive requirements of feminine civility and perfection have in their way fascinated me for as long as I can remember. The stifled rage must've been a constant for women of that era, not knowing that in a few short years everything would change. I can't even imagine how much rage I would've had to stifle back then, because I've had to stifle so much now.

My mother impresses me greatly. I truly think she's one of the most intelligent people I have ever met. Her mind works at a very fast rpm, like a Ferrari brain. She was/is a beautiful woman, and she was preyed upon. Maybe that, besides an agile mind, was what got handed down to me.

But I'm grateful for many other things handed down to me from my family. A dissenting punk spirit. A quick, cruel wit, curiosity, love of history, and above all, a love of words. One of the great things that both my mother and my father gave me is this ability to see art everywhere. I fantasize about having tetrachromacy, where you can see over a million colors. I see shapes and patterns in everything. I'm always surprised that when people grow up in a more traditional way, sometimes they don't seem to be able to see, *to really see*, the things all around them that are pure art. To me, that is what makes the experience of life. It's also something that helped me survive.

For all the flaws of my childhood, I consider myself lucky to have been raised with a European sensibility. We had Italy and its history, its architecture, and its art. I think Europe and older cultures have a different sense of rhythm and time. I find the system, especially the system I now know best, the American system, aggressively determined to crush free thought and those it labels "other." I'm here to tell you that "other" is where it's at.

People tell me they're sorry for how I spent my childhood. That's cool, I simply tell them I'm sorry for how they live. Growing up behind the proverbial white picket fence frankly seems as dangerous to me, and a different kind of cult, the cult of the mainstream. I've known some fucked-up people behind those picket fences. At least with my family it was all right there to see. One of the great benefits of growing up, moving a lot, and continuing to do so as I got older was that I met people who thought differently, and in that way I was raised to view the world from a different perspective. I am grateful for that, if anything.

I was also bequeathed the one thing that runs strongest in my family: a strong urge to destroy oneself. The phoenix that has to rise because life has turned to ashes. My life has ashed itself numerous times, more times than I can count. But goddamn, all those ashes built a beast.

I know I am not alone in this life ashing thing. So many of us seem to have this preternatural ability to rise because we have no other choice. It's something that fascinates me about the human spirit. I think our rising is the bravest thing we can do, and I don't think people give themselves enough credit for it. How many times have we been told we'd be nothing? But we are not nothing, we are phoenixes and we rise. All it takes is some bravery. Turning our lives around is the bravest thing we can do. One step at a time, first we walk, then we run.

One of the things people don't realize about cults is that they're all over: it's not just wild-haired cult leaders. Of course it was extreme in the Children of God when they began advocating sex with children and the selling of women, viewing them as merchandise and property. But when it comes down to it, this mentality wasn't far from what I would later experience in Hollywood and the world at large. At least with Children of God, I knew what I was running from. Hollywood and media messaging was a lot more insidious.

I have patches of memory from the night we escaped the commune. Like a movie scene, it comes in flash images. I remember asking my father where my mother was. No answer. I remember the running. Holding my father's hand. And the green corn-like plants with their hard stalks whipping my small face. The lightning, thunder, and rain raging in the night sky. Sometimes in the movies, it rains to heighten the drama. Well, this drama was heightened. The rain was pouring.

Ironic, then, that after Italy, my father would send me to the perpetually rainy American Pacific Northwest as my next home.

AMERICAN GIRL

It was in the bathroom of the plane taking me to America that I remember first really seeing myself in a mirror. I didn't really know what I was staring at, because I had no attachment to the face I was staring at. I didn't know that once I got off that plane, I would land in a world of *here's what you can't do because you're a girl* and *here's why you're different and fucked because you're a girl*. It was like a pink school uniform for the mind. The first place I was sent was a small town naval base. I went from Florence, Italy, to Gig Harbor, Washington. Or to be more blunt, I essentially went from the cradle of Western civilization to a place with rednecks and jacked-up trucks with big wheels. America was terrifying. Loud. Jarring. I hated the food instantly. I hated how aggressive people were. My brother and I were sent, ahead of my father, to live with my step-grandmother, Dorothy, yet another adult I didn't know but was supposed to attach to. She was a tall brunette cigarette smoker with a big throaty laugh. She *loved* America. She talked about it a lot. My first night in the USA was spent in terror of the bear she told me might eat me in the night. The only thing she knew how to cook was boiled tomatoes, and I cried because I missed the food in Italy.

I was taken to Denny's, a chain restaurant with frankly terrible food, the worst that American cuisine has to offer. They had a big menu with pictures on it. I was so excited to see spaghetti on the menu, I started speaking excitedly in Italian and waving my hands around. When it arrived, it was a gelatinous blob. I stared at it. Picked it up with a fork, and instead of being normal pasta, it stayed together as one unit. There was also a big lake of lukewarm water underneath the spaghetti blob. I just started crying because I knew my life was never going to be the same again. I had landed in a world of Tater Tots and Cheez Whiz, and there was no going back. Fuck.

Everything was different. Not just the food, but the land, the trees, the sounds. It rained all the time in this new place. The cars were so big and so loud. The people were so big and so loud. I had never even seen wooden houses. In Italy all the houses had been made of stone. I had never been around Americans. I had never heard music piped in through loudspeakers. My brother and I huddled together when announcements blared out in the supermarket. We'd never seen fluorescent lights. We'd never seen orange cheese.

Dear America, why is your cheese orange? Who decided: "Let's make this an unnatural shade of orange"? It's completely arbitrary. My brother and I thought it was hilarious. We'd point our fingers and snicker. But the joke was on us. We were stuck there.

My first day in my American school I was made to stand in front of the class and lead them in the Pledge of Allegiance. I didn't know what the Pledge of Allegiance was. I could understand English—I just refused to speak it. I heard the teacher say, "This'll get the Communist out of her." I turned to the teacher and uttered just one word: "*Fascistas.*" Fascists. That's what the Italians were during the war, you dummy, not Communists.

Indeed, it seemed the welcome message was unmistakable: *You're different. We must crush the difference out of you.*

There's a tenacious myth that America glorifies individualism, but trust me, if you are a true individual, you will be persecuted. Schools force-feed you the propaganda version of the world and of history. The bullshit version. So that by the time you graduate you're chanting along with everyone else: "America, hell yes, white men are number one!" Why? Why do you say America is number one? Because if you actually look at the statistics, around the world America is not in fact number one at anything anymore, except maybe obesity, firearm deaths, the death penalty, and incarceration rates. Oh, and of course, military might and our other big export: American film and television.

This is when reactionaries start yelling about how other countries are worse, so why don't I go live there, et cetera, et cetera. My view is why not just be better? Why should we continue to feel superior just because other places are worse? That sounds like bad logic to me. We can just be better by thinking differently. Thinking whatever is different about you must be stripped from you is the WRONG way to approach things. Thinking you must be homogenized for everybody else's comfort level, because God forbid discomfort, is the WRONG way, too. Fuck those ways of thinking. Do not bend yourself to make others feel taller.

When I arrived at school, they said to me, "Stop reading what you're reading. This is what you're allowed to read because you're *X*." "Stop doing what you're doing, girls can't do that." The adults I met were dedicated in their pursuit of beige, not all, but most. Our neighbors had no interest in being intrigued or expanded by an alternative lifestyle or viewpoint. They didn't want to know what else might exist out there in the world. They just wanted to kill it because it was different. I longed for my dad and his strangeness. I needed an antidote, fast.

After a few months, my brother and I were flown to a state called Colorado to reunite with my father. Colorado is one of the most beautiful places on earth. We lived in a little wooden house at the base of a majestic mountain in a mostly hippie community, a town called Evergreen. I loved Colorado, even if I still was not a fan of American food. Soon after, my new stepmother arrived, and life in Colorado was mostly good. I adjusted to my new life better in this freer environment than I had living by the naval base. I had come to love Colorado even if I didn't understand a lot of the social cues of my school peers. At least I was treated well by the teachers, so that was a nice change.

Around this time, I found a book on astral projection. Astral projection is the practice of essentially leaving your body behind and traveling by spirit. I would lie in bed and practice my hardest to get out of my body. I wanted to travel and find my mother.

My mother was still in Italy, and unbeknownst to me was making her way back to America to a state called Oregon. Later I would find out that my dad essentially left her behind to get out of Children of God on her own. Her only living relatives were her sister and her grandmother Vera. Grandma Vera sent her the money to get home and helped my mom restart her life in traditional society.

One day I was told I'd be going to Oregon that night to join my mother. I was excited at first, before I understood that Oregon was not going to be a happy place for me.

Looking back, I have to say, I am incredibly impressed with my mom; she made it back from Italy, raised six kids on no money, occasionally surviving on food stamps, while my dad was living with his new wife. My mom had six kids not because she had really wanted to, but because the cult had encouraged her to. To me it seemed she was embarrassed about her life. Even now, I know she dislikes it when I talk about the cult, but to me it's not her shame, it's just an alternative adventure she went on. There is no shame that should be hers, plus I'm certain it was my father's idea.

When my mother landed back in America, her grandmother helped her get government housing. These houses were pretty basic compared to the prettier home I lived in with my father, but I was ecstatic to be reunited with my mother and other siblings.

Unfortunately, as the oldest girl I got the shaaaaaft. I had to be Mom Jr. I was ten. Taking care of a gang of wild children is not easy when you're a kid. I didn't want to be a substitute mom. I was not suited for it because I like to think too much and get agitated when I can't. I need quiet. I didn't want to be the enforcer, I wanted to go and stare at the clouds. My style of child rearing was not with the best bedside manner, to put it mildly. I was getting angrier and angrier at the circumstances of my life. My powerlessness. I knew I had to help my mom, and I did, but I was not cheerful about it.

Oregon was where I learned to understand the value of a dollar. I discovered what it's like to struggle and be embarrassed when you leave the free food line at church with your block of bright orange cheese. The sadistic school receptionist called our names out over the loudspeaker so everyone in the school could laugh at the poor kids who had to claim free lunch tickets. I'll never forget the sneer on the receptionist's smarmy face when I had to pick up my tickets. Complete classism. I resented those lunch tickets, not to mention the disgusting food. I scalped the tickets on the side to make a little profit. I've always been very entrepreneurial.

Things were going somewhat okay in Oregon, which was to say, hard. Then I met a guy named Lawrence. He lived down the street from us. He had caught me sneaking under his fence to feed his dog bread and give it water. The dog was tied up, chained to a tree. His collar was so deeply embedded in his throat that it had maggots all around it. This was a severely abused animal, which should have given some indication of Lawrence's character. He caught me feeding his dog and threw me out of his yard by the strap on my overalls. I landed on my ass. I hated him instantly. Two weeks later, I got home and who should be sitting in a chair in our living room but Lawrence, with his fat belly, making everybody wait on him hand and foot like he was king of the castle. I walked in the door and he looked at me with a sadistic smile. I froze. He said, "Hi, Rose. Call me Dad." I remember just screaming, "NO!," and running into the field that was behind our house, hiding there. Soon after, he moved in with his two daughters, Autumn and Mary, and his son, Larry (Junior). Lawrence Sr. was charming at first. But I wasn't falling for it. I knew what his dog looked like. I knew the hell this man could bring. He was truly evil, and he had my mom snowed. I kept desperately trying to tell her, but she wouldn't or couldn't listen to me. Despite the fact that she'd escaped Children of God and its patriarchal structure, the societal programming that a man was going to save her was so deeply embedded, she couldn't see the truth. She was probably also lonely. My mom didn't have any girlfriends because she was so busy with so many kids and working full-time.

Probably the first time I ever saw anything sexual was walking into my mom and Lawrence's bedroom. I didn't even understand what was going on, when suddenly a shoe was thrown at me. That was my introduction to sex. A shoe to the head.

As it turned out, my creepy new faux dad was molesting his daughter Mary. We found out years later when she bravely brought charges against him. I always felt something was really wrong, though, instinctively. Mary, who was about fourteen at the time, and I were forced to take baths together while Lawrence watched. Having only recently met her, it was distressing for me. I knew something was off about this and hid myself as much as I could. We would both huddle and turn away.

Apparently Lawrence liked blondes, and thank God I had dark hair. But I saw him looking at my sister Daisy, who was blond. Sometimes she would walk down the hall and I would see him stand up and start following her. I would block his path and get in his face. Well, I would get in his stomach, because I was ten. One particular time I spit at him and it landed perfectly on his lips. He gave me a beating and I took it. I was damned if I was going to let him do anything to Daisy. One of my proudest achievements is keeping her safe from him.

But there was only so much I could do to protect my sisters and brothers during this period. I did my best. My youngest brother, Robby, was three and a half, so cute and so pretty. He rode his Big Wheel tricycle down the street, and for some reason, it set Lawrence off. Lawrence marched out, took spiky rosebush branches, and beat my little brother until he was bleeding all over his back and his bum. Outside of the house, Lawrence made his son and daughter forcibly hold me up and make me look through the window to watch him as he beat my brother. I had never hated anyone more.

Lawrence would find other sick ways to rile me up. He knew I hated the N-word and he would get in my face and say it over and over until I lashed out at him, so then he had an excuse to beat me with his belt buckle. Now I look at ten-year-olds and I think, *Jesus, they're small. I was small.*

Lawrence listened in on the line when we had our weekly phone call with my father to make sure we didn't tell him what was going on. He monitored the mailbox, too, so we couldn't send any letters for help. It drove me insane that this worthless human had power and control over me. The beatings were one thing, but silencing me was his favorite thing and what I hated him for the most. He punished me by not allowing me to speak for a month. I was not allowed to utter a word. I felt so violated. I have had my voice stolen many times since, but that was a big one because it was so literal. I have no idea what I did to get grounded off speaking; knowing me it was probably for talking back. When it was mealtime, I would look across the table at my mother, beseeching her with my eyes, trying to get her to intervene, but there was nothing to be done. Having a voice and being heard is a fundamental human right and this indignity set a kind of pattern in my life.

After a while, my mom did try to break up with him, numerous times. She'd say it was over, that he was gone for good, and I'd get my hopes up, but then he'd be back. One night, he drove around in his pickup truck with a shotgun threatening to kill my mom. I hid in the back of the truck and when I caught sight of her running, I'd shout warnings: "He's coming from the left, he's coming from the right, go right!" Finally, after that, she broke up with him for good. Abusive relationships are no joke and extremely hard to escape, but she did it.

Later, my mother found out he'd done terrible things to all these other women he'd been with before and after her. He'd gotten off every single time in court, playing the system. He even got off the molestation charges. The cops loved him. He got off all the time. No one would listen. Eventually, Lawrence kidnapped his son's girlfriend and raped her across three states in the cab of his truck, holding a shotgun on her. He was finally arrested and sentenced to jail time. I hope he's dead. I hope someone drove a stake through his heart.

I think of how many kids are abused, and how heartbreaking it is that no one helps them. And then it just begets more abuse. I don't know where Lawrence's kids are today. Poor Mary. I hope that poor girl is still alive, and I hope she's doing okay. I hope her and her siblings haven't had their lives totally ruined.

I remember thinking as a young girl, *How is it possible that women can be so gullible? They just ignore the reality of what is happening and believe what they want to believe.* I think women in general, and my mom for sure, got sold this bill of goods, the story that a man will save them. I don't think that's really changed even for girls today. We're still getting sold the same story. I had to unpack it because later even I was ensnared in an abusive relationship.

We need to look at why so many women believe a man is going to save us. It's not because of evidence of saving. I haven't seen a lot of dudes on white stallions pulling up to single women's homes. In fact, I have seen most women get on their own damn stallion. It's just male-dominated society that snows us into not noticing it's we women doing the saving. We are the white stallion and we have to wait for no one but ourselves.

Even though I had these early experiences with men who were horrible beasts, I still somehow got it imprinted in me that a powerful man was going to come along and make my life easier. In reality, they usually just complicated things. Even though rationally I knew it wasn't true, there'd always been this feeling deep inside me that I was bad, and men were, somehow, the superior ones. I was bad because I was tempting. I was bad because there was want attached to me. Lawrence was truly a psychopath, probably the first true psychopath that I met. I would go on to meet others, but he set the mold. There's a direct correlation between my relationship with my father and Lawrence, and later on my relationship with men for the rest of my life.

RUNAWAY THINKER

Everyone thinks Oregon is full of peace-loving hippies. Not the people I was around. They had jacked-up trucks, boosted up with big wheels, and gun racks in the back windows. There were dead deer hanging upside down from practically every carport, with blood draining into a bucket. I have never, in all the places I've been, been in a place more happily vicious than Oregon. I know others have had different experiences there, and I am glad for them, it just wasn't my experience.

People had severe reactions to me there. They went out of their way to tell me I was strange and hideous. I remember encountering a mother in the Fred Meyer department store—she must have been about thirty, a grown woman—who jerked her little girl away from me when I smiled at her, calling me an ugly freak. Her daughter started to cry. I decided to go and see what it was she saw. What was it about me that was ugly? I went into the bathroom and stared at myself in the mirror. My eyes looked even, my nose did, too; I couldn't figure out what it was. I had short hair, but what was wrong with that?

One day when I was about eleven, walking down the street in Santa Clara, a suburb on the outskirts of Eugene, I heard some really awful rock music and a loud car exhaust. I knew this was a bad combination and I was proved right. "Freak!" the guy in the car yelled. I ignored him and kept walking. Next thing I knew I was hit in the head and covered with a brown liquid. Wet from my head down to my toes. The car sped off. As I wiped at my eyes I saw a giant plastic bottle of Pepsi with its top jaggedly cut off. Then I noticed the stink rising. The bottle was the driver's chew spit. It was like the movie *Carrie* where she's doused in blood, except for I was doused in nicotine and saliva mixed with some old soda. I didn't cry, I just sighed and went home to hose myself off. The chew nicotine smell didn't leave me for a week. Every time the air moved around me I could smell the hate.

My father was still living in Colorado at this time, and my parents decided I was to go back to Evergreen and live with him again. We kids bounced back and forth fairly often as there was no formal custody agreement. It was such a strange dichotomy. I went from Oregon, where I was relentlessly deemed hideous and ugly and freakish, to Evergreen, where I was suddenly popular and considered a beauty. This was a strange development. I looked in the mirror again and stared at the same eyes, nose, and mouth, and wondered why before when I had been in another state I'd had things thrown at me, and here I was being worshipped and anointed with instant popularity. I thought about it deeply and came to the conclusion that other human's reactions were useless to me. Ultimately, it allowed me to cancel out what other people thought of me. Later on, when fame came, this deduction probably saved my sanity.

In the meantime, I was handed one more mind fuck on my very first night back at my father's, when I told him what had gone on with Lawrence. He simply said to me, "Well, you made a mistake, you should have sent me a letter from your school." The idea had never occurred to me. That effectively shut that conversation down and made the whole situation somehow my fault.

The two sides of my father became more pronounced. His light side was still magical. He made things fun simply because of how he reacted to the world. My father had a laugh that sounded like this crazed hyena and just when you thought it would stop, it would continue, and everybody else around him would start laughing, too. I can still hear it today. But at this point the dark side was starting to appear more regularly. He was getting angrier and angrier that the little girls in the family were growing up, and not so worshipful. That included his wife. He was having more flashes of rage and becoming more and more cruel. Eventually I had to go back to Oregon to my mom.

A few years later, I was attending Madison Junior High, my least favorite school in my spotty scholastic career. In eighth grade, I went to my first and only school dance. It took place in a squat brown building with bad lighting and cheap decorations. I was skirting around the edges of the room, on the sidelines of the crowd, when I heard a gravelly voice say, "Heyyyyyy. You wanna hallucinate?"

His name was Jack Fufrone Jr. I recognized him from sex-ed class, where we had just learned about fallopian tubes. He had a curly oiled mullet that was strangely mesmerizing, and one of those downy molester mustaches that young rednecks like to cultivate. It was clear my teenage drug dealer had been held back a few grades.

Fufrone Jr. tore off a tiny piece of paper and told me to put it under my tongue. I had no clue what acid was, but I was all in for adventure. He had handed me a tiny corner of a tiny square of paper. I looked at him and took the rest of the square, too. Soon music was pulsating off the rec room walls, and my ears heard every little noise. I left the dance to wander the grounds. Trees started to breathe. My soft young mind was on fire.

After the dance was over, my friend Linda took me home and dropped me on my front lawn, where I lay tripping my brains out, pine needles in my hair, staring up at the trees. My mother came out, dragged me inside, and propped me up on the couch. Furious, she began the interrogation. "So what is it? Are you stoned?" I didn't even know what that meant. "Are you drunk? Are you high?" She kept pummeling me with questions, and I was so annoyed because I just wanted to feel the feelings that I was feeling and see what I was seeing, without this rude interference.

Since the acid had rendered me mute, I had to marshal the strength to speak. I managed to summon just two words: "Fuck" . . . "you." It was like a silent bomb went off. I had never cursed at my mother. Major miscalculation.

By this point, there was another man in the picture, my new stepdad, Steve. He was a mean dry drunk. I remember him telling me that mosquitos never bit him because he had mean blood. He was not at all into us, my mother's children. We could all tell he didn't want us to exist. But we did, so there was a problem.

He was not kind to my younger brothers. Brutal. He didn't like me, either, because I could see him for what he was, and I was always trying to alert my mother. Steve saw his opportunity to get me out of his hair, and he jumped at it. He started in that I was a drug addict, had all the earmarks of a drug addict, because I liked to wear all black and listen to the Doors. One hit of acid. One. Hit. I'm fairly sure it requires more to be an addict.

Two weeks later my mother deposited me in a drug rehab program where I was locked up, at age thirteen, my shoes taken from me to prevent my escape. I told the doctors that I had never taken drugs in my life beyond the one hit of LSD, and they told me I was in denial. Hats off to them: there was no way out of this one. My home for the foreseeable future was the top floor of Sacred Heart hospital, in miserable Eugene.

The time I spent in rehab was both entertaining and monotonous. They taught us about drugs for about four hours a day: what the street names were, what the street value was, where you could get it, what its effects were. Everything you ever wanted to know about drugs but were afraid to ask, straight from the authorities. What the fuck? Did they want repeat customers?

I was by far the youngest person there and soon became the ringleader. One time in the dining room I snorted Sweet'N Low sugar substitute to prove how tough I was and to piss off the nurses. I had never snorted anything, but I saw it in one of the hospital's educational films. That was maybe the most painful thing I've ever shoved up my nose, so the joke was on me. I can honestly say that sugar substitute is a real chemical. You probably shouldn't ingest it and certainly not snort it. The drain was vile. It tasted like rat poison. I managed to keep a poker face and refused to cry like I wanted to do. Making it seem like nothing was probably the best acting I've done to date. The nurses were very unhappy, but I got a cheer from my fellow rehabbers.

There was family therapy one day a week and that was a joke. Everyone in your family had to tell you about how they were affected by your drug use. It didn't really work because I had only done one hit of acid. Mostly my brothers and sisters looked confused. My sister Daisy told me to just say I was a drug addict, that it would go easier on me. Once again, if I admitted to something that wasn't true, it would pacify those in power. If I admitted I was a drug addict, they'd let me out sooner. I thought about it, but no, once again, I refused to betray myself to make my life or other's lives easier.

I was furious to be stuck somewhere and under tight control. The amount of time I was supposed to be in there was dependent on my good behavior. But I knew myself, and I knew my attitude was not going to improve. I was the unit's problem child because I was, quite literally, a child. From what I could see, my behavior was never going to get better. The only way out was to escape. So I did. I had become friendly with our floor's janitor. He did not care at all when I slipped past him into the stairwell and even waved good-bye.

I made it out to the street and just ran. No small feat considering I was only wearing the hospital sock booties with the little gripper pads on the bottom.

I wandered for a few blocks until I came to a coffee shop. I met a girl in the bathroom while piercing my nose with a needle, as one does. She helped me push it all the way through. Her name was Chloe. On the street you meet people and they become your instant best friends. Chloe introduced me to two older punk rockers named Slam and Mayonnaise, total street rat degenerates, both in their late twenties. One had dark hair with big spikes, the other blond with big spikes. There were lots of teen vagrants in downtown Eugene. I was hoping it wouldn't be my future.

It was raining hard that night and we all sought shelter under a church's porch. My bed for my first night as a homeless teen was the cold, wet dirt. It was the oozing mud that woke me up, seeping into my ears. My hearing was distorted, but I could make out some high-pitched screams that didn't make sense. I could vaguely see in the dim light that Slam was on top of Chloe. I didn't see Mr. Mayonnaise. To this day I don't know if it was consensual. I hope it was.

Once again, I think I was left alone because I looked like a boy. I remember feeling saved because I didn't have breasts yet. I slid out of there inch by inch, losing my socks in the process. My ears were killing me, and my vision was starting to double. Barefoot and covered in wet mud, the only thing I could think to do was deposit myself back at the hospital, so that's what I did. I collapsed at a nurse's feet crying about punks and possible rape.

No one believed me. No one would listen. I've never lost the wondering and guilt about Chloe. It is something that drives me to correct injustice.

Everyone was very relieved to see me back, but two weeks and many educational drug movies later, I left again. My roommate gave me some shoes, a couple of sizes too big but better than nothing. I got three other patients to open an alarmed door so I could leave in the elevator.

This time I escaped for good. My life as a runaway had begun.

Being a runaway in Oregon is deeply unpleasant. There's the cold rain, always the rain. Wet jeans clinging to my legs, never fully being dry. And the hunger. I was starving all the time.

There were times when I was a runaway that I woke up after having these weird blackouts. Once I came to while standing on an overpass, woken up because my shoulder bag and backpack had gotten hit off my shoulder and were flying down the road. There were times during difficult moments when I would disappear from my body. While my physical self was left to deal with the repercussions of what was happening, my mind was in another place, gone. That was my method of protection, floating up above, watching everything happen as if through a camera lens. It was not unlike the kind of trances I would go into later while acting, but that wasn't on my agenda just yet.

I had no contact with my family at this point. I was just out there. No one was looking for me. I wasn't offended by the not looking, I was just on my own.

It's funny, on the street you just kind of fall in with other kids like you. The discarded. The uncared for. The lost. One night in front of a Circle K minimart I met a ghostly Nancy Spungen–looking young woman with a mane of fried white-blond hair. She told me her name was Tina and she was a stripper. I had seen a classic film about the burlesque star Gypsy Rose Lee, so I was pretty sure I knew what being a stripper entailed. But when

I asked Tina if she could spin her tassels for me, I was rewarded with a blank stare. She took me to her place, a small box of an apartment with mattresses on the floor and cheap stucco popcorn ceilings. I am not a fan of popcorn ceilings, but I had to make an exception in this case. Kindly, Tina said I could stay for a while. Christmas was coming, and even though I probably wouldn't eat that day, I did want a roof over my head.

After a week, Tina told me I had to put in some money for the heating bill. Aww, damn. What to do? Aha! I decided I was going to rob my mother's house. I made my way back down south to Santa Clara, hitchhiking through small green town after small green town. I finally made it to the house. I waited until I was sure no one was home and crawled in through the cat flap, as I'd done every other time I was locked out.

The house smelled like Christmas. Fuckers.

I picked through the presents, irrationally offended that none were for me. In the movies, the tearstained mom would be on the national news, pleading for her runaway daughter's return. In reality, there was no sign I'd existed. Merry Christmas to me.

I loaded up the wrapped presents satchel style and shimmied out the cat flap. I thumbed a ride in a Datsun 280Z with a guy who looked like Weird Al Yankovic. He dropped me at Pawn-N-Such. I charmed the owner into buying some of my brother's Nintendo games. I got $27, enough for Tina's heating bill.

I was punk as fuck. I've always loved adventures, tiny and large, and this was definitely an adventure.

Tina didn't like me at her apartment while she was out, so at night I would go prowling. Usually I'd try my luck sneaking into gay underground warehouse parties where I became somewhat of a mascot. I would usually do the one hit of acid that I'd manage to procure, or later one line of speed because that was more readily available in the club, and I'd dance until I got kicked out. I would go up on the boxes and just dance like a little machine. That's where I could really lose myself. On the dance floor was where I could channel my fear, stress, everything. I could turn into a kind of dance robot and just move. Sometimes the guys gave me poppers and giggled when I fell over on the floor.

I still favored boys' clothes, and managed from what I could get at Goodwill to put together a Charlie Chaplin–esque outfit: trousers that were a little too short and a bowler-type hat. I took my makeup very seriously. My staples courtesy of Tina were a red Wet n Wild lip pencil, Revlon "Love That Red" lipstick, and Coty white rice powder, which came in a beautiful round box with gold designs that looked like it hadn't changed since the 1950s. I ended up looking like, I don't know, maybe a cross between a geisha and Charlie Chaplin. Some nights I did full checkerboard makeup on my entire face. It would take three to four hours to do a full face of perfect checkers, the black with eyeliner, the white with a stick of pure white Wet n Wild concealer. Sometimes I did spiderwebs on half my face with my eyeliner.

I brought it.

One day I decided to put some acting skills to use. I went to the police station and said I was a runaway, that home was Pocatello, Idaho. I don't know why I picked Pocatello. I think I'd always thought it was a funny name. I told the police that my mother had changed her phone number and I couldn't get ahold of anybody I knew. If they could just buy my ticket to Pocatello, I could find my way. Then I turned on the tears. They didn't know what to do with me. The cops just wanted me to get out of there. They got me a bus ticket. Very kind of them.

There wasn't much to see in Pocatello, so after a couple of days I went to the police there, did the same thing, and they bought me a bus ticket to North Bend, Oregon. North Bend is right on the ocean. I remember being out there on acid, sitting on a log and looking at the ocean. The next thing I knew, the tide had come in. The water got as high as my chest, at the height of winter, before I realized I was practically submerged. The problem with acid is sometimes you lose track of things, like, say, the sea level. I was so cold, my jeans soaked through. Sand went up inside of me. I didn't get dry for, I think, about four days, my clothes stiff and crunchy with sand.

Next I used my acting powers to get me to Las Vegas because I remembered that my friend Lara had moved there. Lara was staying with Bjarney from Norway, who was always trying to get me drunk. After about five days, Lara got in a fight with him and skipped town, leaving me alone with him. He took me to play "blackyak." He pronounced the *j* as a *y*. I got away from him, too.

Pretty soon, I tired of my little game sobbing to the police for bus tickets and crisscrossing the West Coast, although I initially thought it was all very adventurous and adult. I did whatever I wanted, which suited me. But it got to the point where it wasn't so much fun anymore. Being hungry grinds you down. Being cold grinds you down. Shoving down terror and being brave all the time takes a lot of energy. I make it all sound like a madcap adventure, and it kind of was. Until it got old. I woke up one day and realized it was Christmas morning. I hadn't had food in two days. I was over it. At the end of my thirteenth year, I placed a collect call to my aunt Rory in Seattle. I told her I couldn't do it anymore. I was tired, tired of being hungry, being cold. She did the kind thing and took me in—she certainly could have said no. She sent me a ticket and soon I was in Seattle, Washington.

BRUTALITY

My aunt lived in a handsome two-story Craftsman home with her husband, Dean, and her newborn son, Austin. The only people who showed me any kindness during this period of my life were my uncle Dean and Rory for taking me in. Dean was an honorable man from Indiana. He was handsome with green eyes, always a step away from laughter, totally in love with his wife. Dean never looked at me like I was a freak, he never put me down, and he smiled like he meant it. No one else smiled at me, but to be fair, I wasn't smiling much, either. I kind of came in off the streets with a perma-scowl.

My aunt Rory was undoubtedly going through a lot at this time as she'd just had a baby. I really resented that baby. I knew I shouldn't, but I couldn't help myself. He had so many material gifts and so much attention showered on him that it made how I grew up look so lacking. This baby was safe. It made me deeply uncomfortable because I didn't know what to do with these feelings. I was ashamed to be jealous of a baby, but I had never been in a situation where there was extreme care of a child, where attention was paid and love was shown. I craved that kind of love, but if it had been shown, I literally would not have known what to do with it.

My aunt made a big show out of spraying Lysol every time I touched anything. It was humiliating, one more humiliation to add to the long list. To be fair, I had come in off the streets with crabs in my pubic hair and ringworm on my neck, so I suppose Lysol was a good idea. The ringworm looked like someone took a cigarette butt and pushed it deep into my neck, a perfectly cylindrical shape. The crabs I discovered one day while on the toilet. I picked at a little brown dot in my pubic hair and it moved. It had little arms and legs. Holy shit. I almost passed out. When I was taken to the doctor for my neck, he promptly told my aunt about the crabs. So much for confidentiality.

I suppose if crabs and ringworm were the worst I came away with from my stint on the streets, then I wasn't doing so badly. I'd managed to avoid or outwit the predators—the trolls, as we called them on the streets—and not get molested, assaulted, raped, killed. I'd done okay for myself.

One day, about a month after I'd come in from the streets, my aunt told me that my father was moving to Seattle. My stomach clenched. I immediately felt a deep sense of dread. The last time I'd seen him was after he'd moved to Montreal. He was growing more and more unstable. I knew his arrival would spell trouble, and I was right.

I figured he had done something bad to my now ex-stepmom. Why else would he have left with nothing but the shirt on his back, his art supplies piled into a car, hauling across the country like a bat out of hell? Why else would he have deserted my two half sisters? I knew he must be guilty of something major. As he got closer to Seattle, I felt the coming storm.

My dad and I moved into a not-so-charming apartment he called the Cave. It was on the first floor, a dark, dark place with low ceilings, ugly parquet floors, and bad vibes. It must have really depressed my dad, because he'd lived in beautiful houses previously. The building manager lived above us in a hoarder's paradise.

Our apartment was clean, but it was devoid of any kindness, which matched what was going on inside the walls. At this point, my father was deep in a rage he had had against women all his life, but now it had a clear focus: me. He would go off on me, and all women, calling me a "feminazi." He sounded like your average schizophrenic on the street, arguing with some nonexistent entity about women, except I existed corporeally.

I have been dealing with men's hatred of me simply because I am a woman for my entire life, and it all started with my dad. We were born enemies based on gender. His excuse for his rage, for every failure, was women. All women were to blame. Therefore, I was to blame. I came to hate him as he hated me. The worst part was remembering what a magical being he had been when I was little. This monster in his place was the worst kind of betrayal. There are few photos of me that exist during this period, because my father said I was too ugly to photograph. After my years in Oregon, I was used to being called ugly. I would roll my eyes when he said that, but it still stung.

We had no silverware at the Cave, or at least I didn't. I was told I wasn't worth buying silverware for. So I stole utensils from restaurants. I didn't have a bed because I was told I wasn't worth buying a bed for. My bedroom was the closet, where I slept on three pink square seat cushions taken from my aunt's house. I wasn't worth a lot in these days, apparently.

My father often said things like "I can't imagine anybody would ever want to be your friend" or "I can't imagine anybody liking you." He called me a whore almost daily to the point where I'd finish his sentences. I'd verbally mimic him as he went along.

I knew he was wrong. I knew it was bullshit. The thing is, it still sticks. It gets through your walls of defense no matter how high you build them. It grinds you down, hearing this sort of stuff, day after day, being told you're worthless or ugly.

Being a free-spirited, strong-willed, independent young woman (to put it mildly), with a manic-depressive, woman-hating father was exhausting (to put it mildly).

At least I could slide the closet door shut and be peaceful

in the dark. Except I was not at peace. I never knew when he would come home, enraged by God knows what, spittle flying out of his mouth, wild dark eyes that refused to see me as anything other than everything he hated—a representation of all women.

One night the closet door got thrown open. A shaft of light blinded me, but I knew it was my father standing there. He let out a yell and grabbed me by my neck. He dragged me out of the closet and onto the floor. I managed to choke out that I was going to call the cops. He said, "I'll staple your tongue to the floor." I'll never forget the hatred in his eyes, but it wasn't even me he was seeing, it was all women. I knew this, but it didn't make it easier.

Once I tried to tell my aunt what he was doing, but she got mad at me and told me he was the best father she knew. That effectively shut me up. I was stuck with him and I couldn't see a way out. I used to sit in my bedroom/closet and write by flashlight on a yellow legal pad. I would write one thing, over and over, something I called "The Death Monologue." It was a catalog, essentially, of my father's sins and wrongs. My plan was to stand over my father while he lay in the dark on his bed and read it out loud. After delivering my blistering, operatic condemnation, I would then kill him with a meat mallet. Smooth on one side, spiky on the other, with a nice heft to the wooden handle. I was going to beat him to death.

Ironically, the perfectionism that had been ingrained in me by the cult that he'd forced me into probably saved me from spending my life in jail for murder, because I could never get my monologue quite right: each day I had to update the list of his asshole-isms, so the list was never finished. Well, that and the fact I knew my father at this time wasn't worth jail.

One weekend he went out of town. I thought it would be a great idea to have a party, since I didn't know anybody in Seattle and I wanted to make friends. I made flyers with the Cave's address and papered the main drag. Friday at 9:00 p.m. came, and with it a knock on the door. I opened it to see about twenty random weirdos.

In my head, I think I had been expecting a *Breakfast at Tiffany's* party-scene crowd; what I got was Seattle street kids and some creepy adults. The doorbell rang again, now there were twenty more people. Oh shit, it was 9:05 p.m. and it was already getting out of hand. I was starting to regret the whole idea when the police showed up.

Busted. The officers came in, made fun of the generic beer, and broke up the party. I decided to pretend I was just a partygoer myself and walk out with a casual "Good night, Officers." The next thing I knew, I was picked up by my throat and slammed against the wall of the apartment building, a cop's nightstick buckling my knees while he held me up by my hair and smashed the back of my head into the brick wall. My eyes rolled backward and noticed the landlord staring down at me from the balcony above. She was smiling. What a horrible woman. Who could do this to a child? I was maybe five feet tall and barely past my fourteenth birthday.

The cop who beat me looked to be in his fifties, with a handlebar gray mustache and matching slate-colored hair, pasty white skin, and mean blue eyes. My head was cracked open and bleeding, my vision blurred. I was handcuffed and thrown roughly into the back of a cop car.

I was charged with assault on a police officer. The report alleged I had walked up to him and kicked him in the shins. I think besides the cracked head, this made me the angriest. You couldn't give me a better story? So lame. So insulting. I would never assault a police officer by kicking him in the shins. I was just being polite.

When my father came back from his trip, it was World War III between us. Sure, I was guilty of throwing a party like every other stupid teenager in the world, but in no way should I have been beaten by a fifty-year-old male police officer. But to my father I was guilty of everything and more.

When it came time for my sentencing, I was taken in front of six grim-faced men on a high bench. I pleaded my case, but it did no good; they each took a turn pronouncing me guilty. Guilty, guilty, guilty, guilty, guilty, guilty. I was infuriated by the fact that no one would listen to me, by the harsh reality of my powerlessness. They sentenced me to work at a mental institution. The mental institution was outpatient, which meant it was a dangerous situation for a young girl. They either didn't know or didn't care. I was treated like trash and thrown away. It wasn't the first time and it wouldn't be the last. But I'll tell you what, if that same situation had happened to a fourteen-year-old GOC, she'd probably get a worse sentencing. My head took a long time to heal, but my spirit remained unbent.

I started my sentence grudgingly. I mostly did filing, and sometimes they made me mop the bathrooms. When I took my break, a man who lived in the apartment opposite would jerk off out of his open window while staring at me, wiping his semen into shapes on the glass. I later found out this apartment housed a lot of the outpatients. In the end, they didn't seem that much more disgusting than the supposedly sane men who would follow me around on the streets of Seattle.

I remember the exact moment, walking down Tenth Street in Seattle, when I started to see myself through men's eyes. Horns started honking and I heard men yelling. I looked around to see what was going on. There was nothing. Just me. I was what was going on. I was the reason for the noise. Except for it wasn't *really* me that was the reason, it was my body and my face. I had developed a chest almost overnight. I felt embarrassed and ashamed. I felt like I was leading men on just by existing, just by having these appendages.

It got so I couldn't walk a city block without men honking, wagging their lewd tongues through their V-split fingers. That's when a new form of disconnection happened for me. I disconnected from the me that I always knew and became two selves. One self was internal and the other self external, one of me became two.

I didn't enjoy the attention because it made me feel dirty. I wanted to see what I could do about it. I went to the Seattle

Public Library and combed through medical books researching breast reductions. The surgery looked terrifying, not like I could afford it, anyway.

The other half of me asked, *Why does a man's desire supersede my right to dignity? What makes certain men think their perversions are more important than a girl's right to exist as a free human in society?* Eventually I figured out that I needed to be like the sad carriage horses that walk with blinders on the side of their face. Otherwise I was giving my energy away, acknowledging the verbal assaulter's presence. What a pity that so many of us girls and women have our peripheral vision taken away simply because of unwanted attention. I heartily disagree with the whole "boys will be boys" bullshit. No, raise your boys to see girls as humans, not objects.

Around this time, I got my first jobs, both somewhat under the table. The first was at a funeral home. A punk friend of mine was living in the attic of the place and told me they needed help setting up viewings for the families of the dead and that he'd give me $30 a week. Having recently turned fourteen and with slim employment options, I said yes. Working at the funeral home quickly became the best time of my week. In contrast to the rest of my life, I found it peaceful. The dead couldn't harm me. I found it soothing. I was more spooked by the fact that I was not spooked. Sometimes the lights in the viewing room seemed as if they were vibrating and they'd flicker. One of my favorite things to do was adjust the furniture and lighting for the family's viewing pleasure. I would stand by the open casket before the family came and I'd do my best to get a "vibe" from the deceased person and adjust my lights and the decor accordingly. I think that's where my love of lighting came from. Only a couple of times did I feel like the dead were going to come back to life and strangle me, but most of the time it was, honestly, relaxing.

I stole a pack of Jesus cards (the cards the funeral home gave out at the service) and gave them to my father for his birthday. Until the end of his life he bothered me about getting more. I couldn't bring myself to admit that his favorite Christian illustrations with the beautiful gold leafing were stolen and belonged to the dead.

In retrospect, after losing a few people in my own life, I could've lived without some of the things I saw at the funeral home: bodies wrapped in clear blue tinted bags, their legs crossed, rings still on fingers, stacked on shelves in the freezer. One time I opened the window flap of the crematorium. That was not a good idea. Can I honestly say they are all *your* ashes in an urn? Maybe, maybe not.

My other job was at an old movie theater called the Broadway. I loved that job, too, but for different reasons. Two films played for long periods of time when I was working there: *Who Framed Roger Rabbit* and *Working Girl*. Two dichotomies in a sense. One was Jessica Rabbit of the "I'm not bad, I'm just drawn this way" fame, and the other was Melanie Griffith as Tess, a secretary in NYC who dreams of bettering her life and so masquerades as her bitch boss played by Sigourney Weaver.

These two films had a strong influence on my young mind. At home we mostly watched classic films, so newer films were a treat. I was inspired by the idea of rising up like in *Working Girl* and making my way through a male-dominated world to excel beyond what people thought I was capable of. With Jessica Rabbit, I realized that not only were our bodies shaped similarly, she helped me realize I loved to sing. On matinee days when there were one or two people in the theater, I'd go into the darkness of the aisle, put a flashlight under my chin, and sing, "you had plenty money 1922, you let other women make a fool of you . . ." then click the light off and run back into the box office. I considered it my singing time. So I owe my love of singing to a cartoon.

Usually, I was relegated to the front of the house at the theater. The male manager would stick me outside of the big double doors and encourage me to ask men if they'd like to come in and watch a movie—the theater's very own Lolita there to lure men in. It felt dirty and the men who'd come would look me up and down first as if it was their due since they were going to buy a discounted ticket. Later on in Hollywood it would feel the same.

I got fired from both jobs within a month. The funeral home because they found out my age and the theater because I said no when asked if I wanted to be in on the skimming of box office money.

My dad told me we were moving out of the Cave and into a beautiful Craftsman home from 1908 that he got a great deal on. I soon found out why he got a great deal on this place—it was the only house on the street that was not a fraternity house. Yes, I as a girl with dyed jet-black hair, black clothing, and an even blacker expression, was plunked down smack in the middle of University of Washington's Frat Row. You have got to be fucking kidding me. Walking to the bus stop was super fun, let me tell you. It was like running a gauntlet of stupid. The jocks would scream at me as I walked, calling me a freak. I was like, yeah, stupids, I've heard this before and it was boring the first time. Each morning, I'd rig my boom box stereo speakers to my window and put on the band Alien Sex Fiend (imagine a cat screeching along to industrial music) on a loop as loud as I could to annoy them.

My father and I were archenemies at this point. I hated him with every fiber of my being. I wanted him to be in the funeral home dead. He may have wanted the same of me. My father's unbalanced mayhem was daily at this point. My older brother came to live with us, which did not go well for him. My father was mercilessly vicious to both my brother and me, raging even at breakfast. It was such a constant steady stream of annihilating put-downs. My older brother is wickedly funny, but had a very rough time in his teens and dealt with the constant abuse by retreating into himself. Just like he had blond hair and blue eyes, and I was dark, our personalities were night and day. I was tough, he was sensitive. The truth was, I was highly sensitive as well, I just had to have a kind of self-made invisible armor around me to survive.

One night my brother went to hang out with our younger cousin Stevie just outside of Seattle. They were walking down the side of the road when a drunk driver in a pickup truck swerved and killed my cousin, knocking him ten feet out of his shoes. My brother had to see our poor cousin's brain spread across the pavement. When my father found out, he told my brother it should have been him. My father's untreated mental-health issues were creating all kinds of damage at this point. I longed for his old self to come back. I would watch kids with their dads and I would get jealous.

It came time to enroll me in school, but I didn't really have the necessary documents. We found an alternative school called Nova that catered to kids too smart or too weird to be anywhere else. It would've been a good place for me, except for it was all hippie kids. I fucking hated hippies. They reminded me of my earlier years and they pissed me off because I equated them to the hippies in the Children of God. The hippie chicks would lecture me about all the chemicals in my Revlon lipstick. I would respond by whipping it out, applying more, and blowing a kiss. I also enjoyed leaving kiss marks on the white walls. I always got in trouble; it was obvious who was doing the kissing.

My dad took me back-to-school shopping at REI, an outdoorsy I-shoot-bears kind of store with cargo pants and anoraks—not my idea of where to buy clothing. I was more of a Moth, a cross between a Mod, that sharp 1960s look, and a Goth. I fashioned miniskirts out of black wide-necked turtlenecks from Goodwill, cutting off just the neck part and wearing it over black tights, with a pair of Doc Marten lace-up boots that my aunt bought me. I had my own weird flavor going on.

When I would get any little amount of money, I would spend it on one hit of acid or a line of speed and go dancing at an all-ages club called the Underground on Friday night. I loved this place, lots of Goths and freaks. My kind of people. Not only did they play fiercely awesome music, but I, higher than high, would rule the nightclub stage, like I had when I was a runaway on the streets. I'd get up there, extend my arms, knock the queens off, and basically announce, "I'm here, bitches, step aside." I was there dancing like a little machine till 2:00 a.m. I'd usually fib and tell my dad I was spending the night at a girlfriend's house, which created a time problem for me. I had nowhere to go after two in the morning, so what I decided to do was to go sneak in and sleep in the Lake View Cemetery, one of Seattle's oldest cemeteries. The cemetery was gorgeous and Bruce Lee was buried there, his grave usually covered with oranges and pennies. I found the cemetery's silence and safety reassuring. Plus, weird sex stuff was going on in the bushes in the park next door and I wanted no part of that. Come Saturday morning I'd straggle home, my arms filled with flowers that I'd taken from various graves. When my father asked where I'd gotten them from, I'd simply say a nice man gave them to me on the street. Amazingly, he bought it every time.

Other nights I'd just sneak out, making a big show of going to sleep in my closet and then leaving out the front door when I could hear my dad's soft snores. I'd make sure to leave the front door open in case anyone wanted to come in and murder him, but no one ever did, much to my disappointment.

When my dad announced I "owed" him $300 a month in rent, it put the fear of God in me, or at least the fear of being forced back into homelessness. The experience of being homeless and hungry casts a long, long shadow. I'm not sure you can entirely imagine what that's like unless you have been hungry, really hungry, at some point. Can't-sleep-because-your-stomach-is-eating-itself-for-the-third-day-straight hunger. My fear of landing back on the streets was like a blackness looming behind me. I was terrified of living like that again.

But how on earth was I going to come up with $300? The answer came when I was skipping school one day. I saw a flyer on a pole that said I could get $35 a day to be an extra in a movie. I'd been poring over classic films, watching with my dad, one of the things that even during this period we still could enjoy together. I calculated $35 a day for who knew how long and figured I could get the $300 that way.

This wasn't quite what you'd call a classic film: a movie called *Class of 1999*, with Malcolm McDowell and Pam Grier, a silly B movie about a future school where the teachers were cyborgs. But hey, $35. I quickly applied for and got the part because they liked my look. An extra is what it sounds like, an extra body, an extra face, to fill in the movie. I was more of a featured extra; I'd just pop up in shots, completely out of place, but because I was cool looking they put me in all over. They drew a little black heart on my round cheek because I was in a gang called the Black Hearts. It was almost like I was a silent film character, all over the film, but never speaking.

There was a guy on set who was probably in his late forties. He was super friendly and he reminded me of a nice version of my dad, so I'd joke with him a lot. I always thought adults liked me because I was so much different from other kids my age, and I was funny. One weekend, he asked me to walk around downtown with him and some of the other extras. I said I would. We were supposed to meet him in the hotel lobby. I waited and waited. I had the front desk call his room and they told me to go up. I went expecting to see my Black Heart extra gang, but no, it was just me. The door opened and I got pulled in right into his chest. Awww, fuck. His beard scratched me as he jammed his tongue down my throat. It all happened so fast. He promptly pulled down my shirt and fondled my breasts. Of course, it was me who felt dirty and ashamed.

Looking back, this man was just another industry pedophile, no different than a street troll, but I didn't know that yet. There are so many of them. It's an open secret in good old Hollywood. When charges are filed (rarely), the studios just continually settle with the victims, and use their PR machines to invalidate the claims. Usually where there is smoke there is an inferno, especially as Hollywood is concerned. I tucked the experience into one of my many inside compartments and went back to work. It didn't occur to me to say anything. For years I thought of the incident as a sexual experience versus assault. Later when I became an adult, I realized that it actually was assault.

The truth of it is, the shame was not mine, and for all victims in similar situations, it is not ours. The shame is reserved for every creep who has ever touched us inappropriately. The shame is on the abuser, not the victim, not the survivor. It is tragic that so many of us have to survive this kind of crap, and I'm so sorry if it has happened to you.

The good thing to come out of my foray into movies was meeting a cool kid named Joshua Miller, a child actor from a famous cult film called *River's Edge* with Keanu Reeves. *Class of 1999* was his first movie without his acting-manager mother on the set; he was thirteen to my fourteen. I corrupted him as best I could. I acted super worldly, and I suppose, in comparison, in some ways I was. At least in the wild and free way I lived. We snuck into a very adult bar called the Pink Door and got tipsy on blue martinis while we watched a cabaret performance. I also got him stoned on weed using an apple as a pipe.

Joshua would go on to become a very big writer in Hollywood, and he and his writing partner, Mark Fortin, wrote *Dawn*, my directorial debut. We developed a lifelong friendship, my longest relationship with another human, besides my family.

It is thanks to Joshua that I ended up in Los Angeles. When his mom came to the set of *Class of 1999* one day and met me, she thought I had the goods to be a star and sent me to Hollywood on the Amtrak train.

This is how I wound up getting my first speaking role, one line in a Pauly Shore film called *Encino Man*. Oh dear is right. I thought the executives and the director and all the studio people on the set were definitely not as cool as they thought they were. I was introduced to an old agent lady named Beverly and she told me I had to change my name because I sounded like an Irish scrub girl. I told her that her last name didn't sound so pretty, either, and maybe she should consider changing it. I had no problem talking back, but the comment about the name stuck with me even though I didn't entirely agree. I toyed with changing my name to Rose Mayfair because I was reading one of Anne Rice's vampire books at the time and there was a character with that last name. Luckily there was a supermarket chain called Mayfair and that discouraged me from changing it. My last name may not be the prettiest, but I've always liked how Rose McGowan looks in print; to me it looks strong.

I was told by the mean agent lady that I wouldn't get hired if I wasn't emancipated. She told me that lots of kids had their money stolen by their parents and while I couldn't conceive of that happening—I believe I had twenty-five cents to my name—I figured it was best to protect myself.

So the next logical step in my illogical life was divorcing my parents, aka emancipation of a minor youth. I was fifteen. I needed freedom like I need air.

I had been in Los Angeles for a few months. The family court building is where you get to see behind the rich LA façade. There were families of all kinds with their bored kids running up and down the halls—terrified-looking mothers, some fathers, and then me, the kid who was representing herself in court. I thought it best to dress like what I thought an adult dressed like, which to me meant tan pantyhose, the kind you get in the drugstore out of those funny-shaped eggs. I was pretty sure I never wanted to look like a grown woman if I had to wear pantyhose, but I could handle it for half a day if only to impress a judge with my adultness. I remembered Melanie Griffith's pantyhose in that movie *Working Girl* and decided that must be what an adult woman wears. I was going to, too.

My name was called. Showtime. My heart racing and my pantyhose falling, I tried to inconspicuously yank them up but only succeeded in making a huge tear. The judge was stone faced. I launched into my case. He tried to point out that a child's place was in the home. Not my home, I said. Besides, only in the technical sense could one say I'd ever been a "child."

"Let me tell you about my home with my father, Judge. The last solid year we spent together. We lived in a grim, dreary place with zero light in depressing Seattle, Washington. My father was an artist. For him to live where there was no light must've been terribly difficult on his already crossed brain wirings. Artists see light in ways others don't; they literally depend upon it. Maybe that's what was wrong with him. Maybe that's why he hated me." But I didn't say all that, I just highlighted the abuse.

It was a matter of survival. I said I could no longer tolerate being a man's property. The judge cleared his throat and declared my freedom. I was technically now an adult, but still not old enough to drive. I was officially no longer the responsibility of my family, which was ironic, considering that they'd never been responsible when it came to me, anyway. And I was emancipated, which is ironic, given how I would spend most of the next few years, being owned by another man. I escaped the control of the cult and my father, only to run headfirst into another kind of ownership.

CAPTIVITY

After the emancipation, I got word that my mom had found work in Los Angeles and was getting a divorce, so she was single—my favorite version of her. By this time she had graduated from the University of Oregon, Phi Beta Kappa with a double major in English and journalism. We were no longer legally bound to each other, and we felt like equals. More friends than mother/daughter. We moved into a great little place, a white wooden house from 1918, right in the heart of super-sketchy Hollywood. It was so much fun.

I enrolled as a sophomore at the world-famous Hollywood High, where I was in the magnet program for artists. I was happy to be in the magnet program because there weren't that many people in it and the kids were all creative. Dancers, singers, actors . . . it was every kind. But after all my more adult adventures, being back in school with young teens was odd.

I felt ancient comparatively even though I was only in tenth grade. In the theater program, they were gearing up for staging a Greek tragedy. I auditioned for and performed my one and only serious play: *Antigone*. Despite being a rookie, I felt like I'd trained for it my whole life. I remember just channeling, vibrating, leaving my body, but this time for something good, a pure energy coming through my body instead of leaving due to trauma. This acting thing was intense and kind of fun.

Afterward two men in the audience came up to me and said I'd made them cry during my death scene. It made me feel powerful. Theatrical acting is very different from film acting, I suppose a purer form of acting in some ways. Hollywood being such a movie town, I never felt that power again during performing. But at the time and for many years after, the idea of doing the same thing day after day, night after night just seemed too much for me, so I never really did theater.

I was truly happy living with my mom. That had been my dream growing up, that it would be the two of us, and that she would finally see that I was worth more than the men who'd come into her life. It was so infuriating that these stupid, worthless men ruled her and thus lorded over me, just because they had something between their legs. I could see that so many of them were shysters, abusers, molesters—why couldn't she?

My mother didn't realize the con jobs that were pulled off by men making her feel like *she* was the lucky one. Like many women, she was taught that winning a man was her ultimate goal. When we are trained from a young, young age, when we are steeped in that, we don't learn to see our own worth. We don't realize that it is us all along who are worth the gold; we are told that we have to settle for silver.

Only later as I got into my own fucked-up relationships would I begin to have more compassion for my mother. I understood then how easy it is to be manipulated and how they can and do prey on you when you're at your weakest. We'd both been taught that men were more valuable than we were. Men were my mom's Achilles' heel; I guess they became mine, too.

One night I went to a famous Hollywood diner, Canter's, and met a guy named William in the parking lot. He was maybe twenty or twenty-one—it was always older guys going for me—and we started hanging out. I think I was most likely just entertaining myself. I don't know what I was doing exactly; I was just hanging out with this guy who liked me and who was kind of cute, and who had a sweetness to him, despite being what I deemed "very Beverly Hills." I lied and said I was seventeen. I think I told him the truth after a week. It didn't seem to throw him. A twenty-year-old with a fifteen-year-old is creepy, but I guess I wasn't what you'd call a normal teenager. Even so, in the back of my mind it weirded me out, but I pushed it down.

Around six months into living with my mom, she met a guy named Stewart. He was Dutch, and his teeth had this brown tinge to them from smoking unfiltered cigarettes. He wasn't one of the worst ones. He was just a pathological liar, but she didn't know that yet. One day shortly after meeting Stewart, my mother sat me down with William in the living room and said, "William, do you care for my daughter?" He said, "Yes, very much." I remember being somewhat confused as to why she was asking this. Then she said, "I'm moving away with Stewart. Will you take care of my daughter?" I was stunned. I had only known William for three weeks.

And that's how I wound up living in a beautiful 1920s duplex modeled after a French château in a leafy section of Los Angeles.

William was a spoiled brat, a Beverly Hills mama's boy. He had gone to Beverly Hills High, which was very famous, very 90210. He was a Hollywood rich kid who refused to get a job, but it didn't matter because his parents gave him everything. Every week, he visited Daddy's accountants, and they'd cut him a check just for existing. Infantilized by his mom, financed by his dad, he was the quintessential spoiled rich kid. I was really resentful and jealous of him that he never had known struggle like I had. But looking back, it's its own trap, just a softer trap.

More and more, I fell into a depression. But I couldn't leave and live with my mom, and my dad was NOT an option. I was stuck, so I had to pretend I liked him. Sometimes I did like him fine, but more as an odd roommate. Really we were both kids playing adults.

I had dropped out of Hollywood High at this point. William kept me like a bird in a cage. He was desperate to control something, and there I was, this young thing that got a lot of attention. I didn't have a car, and you can't do much of anything in LA without one. No car, no work. So I had no money. I would steal a dollar here and there when William left tips at restaurants so I wouldn't have to ask him for tampon money.

He was also incredibly, stupidly jealous. I don't know how I could have possibly cheated on him because I was dependent on him for rides, for food, for shelter . . . but regardless, he would fly off the handle if I so much as looked at a waiter, so I learned how to act, trying to keep him calm. And THAT is how you fall into an abusive relationship: when you start acting in a way so as not to upset the other person or set them off. You've given away control of your own life, bit by bit, bit by bit. It's incremental, until one day, you have hidden so much of yourself you get lost.

If a man flies off the handle over even the mention of a crush or an ex or whatever, it is a red alert. Get away. Fast. The possessiveness only gets worse. Cut your losses and bail.

I didn't. One day I got out of bed and William looked at me and said, "Stop." I stopped. "Stand still." I stood still. "Now turn around." I turned in a circle wondering what was going on. "What are those upside-down triangles on top of your legs?" What? I went to look in a full-length mirror and then I saw what he meant. Or thought I did. What happened in that moment is something that happens to so many girls. I stopped seeing myself through my own eyes. I was now seeing myself through William's sick eyes. Suddenly his version of my triangular-thighed reality was my truth. Snap your fingers. It happens that quickly.

Once again, a man was telling me I was imperfect. Now, had I been a grown woman at this point I would've simply pointed out that those triangles were where my legs met my hips, but even though I was living as an adult, I was technically still a child. When life is big and scary, the only control a girl sometimes has in this world is her food and body. Many people mistakenly think eating disorders are about vanity. Trust me, there's no vanity involved when you're growing fuzzy hair on your whole body because it's trying to insulate you from your starvation, there's nothing vain about having vomit and stink coat your fingers after puking, there's no vanity when you're so full from a binge that you can't breathe. Anorexia and bulimia start out about weight loss usually, but it is about far more than the body. It is the mind. Fear of imperfection in this fucked-up society becomes an obsession-run riot. So I did what I was programmed to do in our perfection-seeking cult. With every movie I'd seen, with every magazine I saw, I knew the tricks and the rules already. I'd been primed. That's the fuck of it all. The programming begins from the day we are born. It lies in wait for you, and then when you're down—that's when it strikes hard.

My legs went from being a part of my functioning body to being the most hated part of my body. Suddenly all I could see were legs made of upside-down triangles. They needed to be controlled! They needed to be destroyed! How had I not seen these grotesqueries?!

I went after those thigh triangles with a vengeance. William bought me a Step aerobics bench to exercise on. I got a Cher workout tape, played it over and over and over, four to six hours a day, sound off on the tape, Nirvana playing on repeat as I went up and down, around and around. I was obsessed.

William would buy me *Marie Claire*, *Glamour*, *Vogue* for "thinspiration." When I was trying to figure out how to best stay thin, these magazines were a great help. I cut out so many pictures, anyone with super-skinny legs. I would sit on the toilet because of the laxatives I was taking and cut out pictures of girls who were thinner than me. And I'd cry. And I'd be mad. Now and then, whispered words would try to puncture through my haze. *This is not who you are. This is not what you care about. This is not your life.* But it was my life. And I was now trapped physically and mentally.

About once every three days I allowed myself to eat something, usually a big pot of pasta.

I never was able to get below ninety-two pounds. For some reason that was my cutoff point. Because I had read about girls who were eighty-four pounds, I felt like a failure.

I was exhausted all the time. It took every ounce of my energy to work out. But I couldn't have an imperfection. Whenever William would start harassing me or yelling at me about who knows what, I'd fall asleep almost instantaneously, chin on chest. It became a defense mechanism. The only time I was active was to work out; the rest of my time was spent trying to check out mentally. It is an infinitely lonely disorder. No one can get to you, least of all yourself. If your life becomes about all these invented perfection "rules," you are no longer in *your* mind, you only hear the evil outside voices. Those evil outside voices become inside voices. And those voices are mean. Kind of like nasty message board/comment sections but in your own head at all times. Super-fun stuff, right?

To deal with it all, I just checked out by going into Step Reebok Land. I lived an impersonation of an adult life with a wealthy son of a Hollywood somebody. I became sixteen, seventeen, eighteen, and then nineteen. I mostly didn't speak for almost three and a half years, which is hard to believe if you know me. But I was too tired. I was a young woman trapped in a strange, empty life.

One year I went to visit my family for Christmas in Seattle. I arrived through the front door, ran through the house and out the back door, and then went running in the snow. It was about ten o'clock at night, and I ran around the lake, which is about three miles. I had my Step Reebok with me, too, of course—couldn't leave home without it! In fact, I had an insane theory that if it wasn't disguised, the baggage handlers were going to steal it for their overweight wives.

It got so bad I was hallucinating. I tried never to sit down. I was sure people could see fat dripping off me.

Starving made me feel a fuzzed kind of high. I remember thinking that at least I was superior to heroin addicts because I was high for free. Yeah, that's how you want to go around gathering your self-esteem, from your superiority to heroin addicts.

But I also think it was my way of surviving trauma. I could deal with starving myself in the name of perfection as self-protection. I could deal with having to tell William I loved him because I'd cross my fingers and hide them behind my back. I had started to really hate him, but I pushed it down because I knew I had no alternative living arrangements.

Thankfully, William and I didn't have sex very often. It felt like sleeping with a half brother, if your half brother had screaming, jealous rages. It was weird. I hated it. Tears would leak down the sides of my face while he'd be pumping away on top of me, thick beads of his sweat falling into and burning my eyes. Sometimes I'd turn to the side and cry, but it didn't stop him. I just went back to my old trick, separated from my body, looking down on myself from above. I had decided to act my way through sex. I didn't know what an orgasm was supposed to feel like, despite the stupid magazines always talking about them, but I figured out that if I faked it well, I could make the sex stop faster.

That was me, an imitation of life.

For years afterward I'd get down on myself for living with William. I blamed myself as if I'd been an adult making grown-up choices, but I was just a kid. A kid that was so, so scared of being homeless again.

In hindsight, I was surrendered to a guy I barely knew. But even though I was abandoned and held a lot of anger about that, I now realize that growing up how I did, you go where the love is, and if the only love you're getting is coming from men who say they want you, then of course a girl's going to go to that. It sets you up for failure. It sets you up for loss of life. It's something that I think people don't talk about in the cycle of abuse.

While I was busy whittling my nonexistent fat away, William's mom died suddenly. I'd say within three days William lost his marbles. The problem with making your children completely dependent on you is that if you leave them early, they are fucked. And he was fucked. It was very sad.

We immediately moved into her big house in Bel Air. Our uneasy alliance went deep south at this point. William was rudderless and absolutely unqualified to know how to take care of himself without his mother. I really did feel bad for him. But as he turned to more and more extreme drugs to suppress his pain, he became erratic. One moment speedily talking and then crashing into sleep for twenty-four hours. I had to walk on eggshells. The occasional kindness he showed me, the occasional sweetness in his eyes, disappeared entirely into drug use. Smoking, snorting, pill popping, whatever. He started disappearing for a day, then a night, then three or four days at a time, leaving me with no money and no food. Since I was anorexic, that was kind of okay.

During this period, I still had to manage the housekeepers and the gardeners, making sure they did a good job. Even though I was living in a fancy house, they had far more money than I did. Since I was busy being anorexic, I had other things on my mind, but the truth was I was stuck in a huge Beverly

Hills mansion that had become a prison. The house was in a canyon, and there was nowhere to go. I was alone.

William needed more and more money for drugs—I think he'd started freebasing cocaine or meth—and he started coming home and demanding money from me, forgetting that I had no money of my own.

I would call William's dad, a big-shot senior executive, and beg him for money to get out, to go anywhere. I asked for $500 because I thought that was a lot of money. But the dad's answer was always a big fat NO. Through one of William's mom's friends, I managed to get a little work doing a small commercial for Allstate, the insurance company, but it wasn't enough to get an apartment or anything, really. Plus, when I cashed the check at a check-cashing place, William took the money out of my hiding place and spent it on drugs. Meanwhile the dad bought William a brand-new Ford Explorer. Because that's appropriate. Reward your son who's a major drug addict with a new car. Hollywood parenting.

One night I woke up with hands around my neck. I screamed and in the low light I saw that it was William squeezing my throat. His eyes were black. No one was home. I made a deep choking sound, and it snapped him out of his fog. He backed up a few steps and looked at his hands. I kept thinking that this was all some ridiculous movie I'd gotten stuck in. He pushed me onto the floor and dragged me by the back of my collar. I was screaming every curse word I knew at him. I was scared for my life. He pulled me outside, across the patio, and

I tripped while he continued to drag me across stone pavers, tearing two of my toenails off. I screamed and startled him into letting me go. The pain was intensely sharp, but I didn't feel it much because I was furious. I got up and limped/ran inside to the laundry room. I came out and *crack!* I smacked William across the head with the broomstick. The doorbell rang. William went to answer it. I heard his father's subdued British tones. Surely he'd give me enough money to leave this time. William, only holding the door open a crack and with me crying in the background, was telling his dad I was upset about a friend. I started yelling. While William was trying to hide the evidence of my blood trail, I called out to his father. "Please, I just need to leave. Please, just help me get out of here!" Instead, the father and William shut the heavy front door on me and left me there bleeding as they drove away.

My feet were on fire and feeling like I had individual heartbeats in my throbbing toes.

I sat on the couch trying to decide what to do.

I couldn't figure out how I was going to get the money to escape, but it was time to go. Then I got a call from my sixteen-year-old sister, Eve, telling me she was running away from my dad and needed to stay with me in California. I knew it was bad with my father, but I also knew it was bad with William. I figured my situation was the lesser of two evils for my sister. When she told me she was arriving at noon the next day, I panicked. How was I going to get her from the airport to the house, with no money to my name?

I decided the best thing to do was to pawn one of the televisions. I, at ninety-two pounds, carried a giant TV down the hill and into a taxi that I had take me to Beverly Hills Pawn. I sold it for $80, enough to get Eve out of the LAX parking lot with $6 left over. My poor sweet sister was going to be met by me, a cranky anorexic with no extra energy to waste on niceties and talking sweetly. It makes me cry thinking of how mean I was to Eve, especially because she was so grateful for me taking her in. And what was I taking her into? A beautiful house with no food and a crazed, unpredictable "boyfriend." Still, she thought it better than where she'd been, which breaks my heart.

William disappeared almost entirely at this point. The gardeners and the maids, paid for by his father, still gardened and cleaned. I thought it was absurd that Eve (and less importantly I) had no food but the house showed to perfection. One night soon after Eve arrived, William reappeared at 5:00 a.m. and woke us up by banging away on his mother's Steinway grand piano. Then he came and started a knockdown fight with me. Eve saw it all. I was mortified. William passed out finally.

After that, every couple of days William came back and slept for at least twenty-four hours, so I knew I had a chunk of time free of him, and I would "borrow" his car while he slept, taking my sister dancing at a nightclub in Hollywood called Dragonfly. At the time, it was a total incubator for talent. I saw Fiona Apple play there before her first album came out, and Mazzy

Star. The crowd was so beautiful. We were all so beautiful then. I became friendly with a lot of people who worked there. The kindness of strangers was a relief.

My mother and I during this time had very little contact, and I just felt like, how am I going to explain this insanity? And I was done with my father, done with the brutality. I just surrendered, thinking, *There's literally nowhere else for me to go.*

That's when I met Brett Cantor, the Pied Piper of People, aka the Mayor of Dragonfly. He co-owned the club as well. Brett had blue-blue eyes and short, platinum, shaved hair. He was lovely. Funny as fuck. He was twenty-five (I was just nineteen by now) and he was the youngest music exec at Chrysalis Records.

When we first started going out, I was too shy to speak; I was so exhausted from being anorexic my talkative side had left me. I had my sister Eve speak for me, if you can believe it. But he was perceptive and figured out what was going on with me. He was in AA. He had been sober for seven years, since he was eighteen, when he'd hit bottom and sold his dad's Irish setter on the side of a freeway to get money for crack. He suggested I try OA, Overeaters Anonymous. Oh, and he got the Irish setter back to his dad. After that, Brett and I became much closer and I started to regain my powers of speech. He showed me such kindness, a kindness that I'd been missing. I started seeing flashes of the girl I really was and the woman

I was going to become. One day, I opened up to him about William and he was determined to help me get out of there. After discussing my limited options with Brett, we made arrangements for me to live with a friend of his. But I needed money. Shit.

Eve and I went back to the fancy Beverly Hills house and looked around for what we could sell, but everything was too heavy for us to carry down to the pawnshop. And then I saw it. The Steinway grand piano! At this time, there was a paper that you could advertise in for free called *The Recycler*. I called and placed an ad to run the next day. William was out on one of his benders and I hoped he didn't come back. I listed the Steinway for $1,500. I didn't know that these kind of pianos were going for $20k; I thought $1,500 was a ton of money, bless my heart.

The next day the house was swarmed with piano buyers. I realized that I could've gotten a lot more money for it, but oh well. It was taken away that day. I pulled a reclining chair over to put in place of the absent piano and went to sleep. When I woke up, William was sitting in the chair. Eve and I were terrified, sitting on the couch staring at him with frozen smiles. He simply sat, drugged out, and didn't seem to notice the missing Steinway. He started talking about how he wanted us to start over in our relationship. I was so nervous, I just played along with him. I told him that it would be best if someone could pick him up and give me time to think about it. Now, it made no logical sense to have him picked up since he had his new

Ford Explorer, but luckily his addled mind went along with the plan. A druggie friend of his picked him up.

I said to Eve, "This is it, this is our chance." We threw everything we owned, mostly just clothes, into the back of the Explorer and took off like bats out of hell. I drove like a possessed person and we made a twenty-four-hour drive in just eighteen hours. Once I got to Seattle, I dropped Eve off and went to the Ford dealership. I forged William's signature on the car's "pink slip" and traded it in for a spaceship-like sports car and never looked back. That was the end of William for me. I felt some guilt, but my missing toenails reminded me that I shouldn't feel too bad.

I was nervous to see my dad, who was frosty to me, but he still gave me an awkward hug upon seeing me. I said a teary good-bye to Eve, got some of my stuff from my dad's house, threw it in my new space pod of a car, and took off. It was time to move on to my new life and that included Brett.

I drove back down to Los Angeles, drinking Diet Pepsi in place of food to keep me going. Not obsessively exercising was really hard; I felt extremely guilty and extremely fat.

I called Brett numerous times on the drive back down, but I couldn't reach him. I kept leaving messages. When I got to Hollywood's Highland Avenue exit, right by the Hollywood Bowl, I called him again from a pay phone. I was starting to panic about having no place to stay. I hung up and called again. A man answered the phone and it wasn't Brett. The voice identified himself as LAPD and said, "Brett has been murdered."

My blood ran cold and I remember nothing after that. I found out later Brett was stabbed twenty-three times and almost decapitated. My world, my hope, went black. I fell to the ground and went catatonic. I don't remember much until the funeral. Everything went so dark and I couldn't stop crying. He had been stolen. I couldn't stop thinking about how much fear he must've experienced in his final moments on earth. How much rage and terror his kind soul had absorbed. I shudder even now thinking about it. He'll always have a piece of my heart. The case is still unsolved, but I have been trying for years to remedy that.

Embarrassingly, I also had the fleeting and fucked-up thought that the magnitude of his death would cure my disordered mind and get me back to being the brave girl I was. I was so bored with thinking about my body.

At the funeral they played "Wish You Were Here" by Pink Floyd and I can honestly say I wish Brett were still here. He deserved to have a full life; he deserved to keep shining. I sat down and cried and cried.

I saw his brother Cliff at the funeral and thought he was the darkness to Brett's lightness, both physically and in personality. It's bizarre, but within two weeks of the funeral, I went to coffee with Cliff. They say statistically there's a high likelihood that the surviving partner will go out with the best friend or relative of a deceased partner. I wound up going out with Cliff, which was a mind fuck for sure. When someone dies, it stops your relationship exactly where it was, so I felt like I went from zero to thirty with Brett, and then went from thirty to one hundred with his brother. It was weird, but men were my ports in a storm, you know?

The friend of Brett's I was supposed to stay with disappeared after I stayed there for a couple of weeks, and once again, I had nowhere else to go. So I wound up staying with Cliff. Cliff inherited his brother's record company job, his nightclub, his girlfriend, and his car. That must've been so psychologically challenging. I know it was for me. I'd look at him and imagine he was my lost love. We both needed each other as a connection to Brett.

I started Overeaters Anonymous more seriously because Cliff was also sober and in AA like Brett was. It helped me immensely. I was called the toast girl, because the first year in recovery I ate toast three times a day. That was what I committed to doing, and I cried every time I did it, because I felt my body expanding and my brain going crazy with the fatness. I kept a diary at the time that I filled with horrible, crazy, mean scribbles like "fucking fat ass" written a thousand times. I guess I was vomiting on the page versus actual vomiting, which I was never good at.

It was after Brett died and after I joined OA that I started doing the real work of my life: dissecting things, studying them from all angles. OA's twelve steps are basically the same as AA's. Granted, I never made it past ten. Those final two steps are explicitly about regular prayer and spiritual awakening and I didn't want anything to do with God. It was too loaded with trauma for me to get there at the time.

Three weeks after Brett was killed, I was standing by the front of a gym, waiting for my friend Josh, weeping over Brett again. At this point the sobs had turned into a sort of constant leak, so I was just standing there leaking, when this woman came up to me and asked me if I was an actress.

If she had been a man, maybe I wouldn't be sitting here right now, because if you're a beautiful girl in LA, countless men come up and say, "Are you an actress?" just to see if they can sleep with you. But because it was a woman I gave it a little bit of credence. I told her no.

"Well, do you want to be an actress?" No.

I had just turned twenty. My idea of actors was based on the stereotypes, that actors were selfish, self-obsessed, fame hungry. I had the self-obsession part down because of my disorder, but not the other parts. I never had that hole in my soul that could only be filled with outside things, including fame. I knew from William that "things" couldn't make me happy. I knew deep down that happiness was an inside job. It turned out the woman was the producer friend of a director named Gregg Araki, a famed indie director I hadn't heard of. Gregg had looked far and wide for the lead in his upcoming film *The Doom Generation* and hadn't found her yet. I was undecided, but I gave the woman my number. They called for about a week trying to get ahold of me. It was my friend Joshua from *Class of 1999* who convinced me to go meet with the movie people.

At last I met with them. I had to go deep into the Valley, outside of LA. I remember not wanting to drive all that way.

But after asking how much money I'd get paid ($10k), I realized it was enough for first, last, and deposit to rent a room in a house. I'd have enough left over to go to Paris and visit the empress Josephine's summer home, the Château de Malmaison. I would not have to go back to cold, rainy Seattle and my dad. My deciding factor on being an actress came down to that: Dad, Paris, and rain avoidance.

So I met the director; I thought he had a fun, infectious personality. The audition commenced. One of the male stars was there. Araki told me they needed to test us for chemistry. The actor was lying flat on a couch. I was made to lie on top of him. He was lying on his back and he had an erection. I could feel it. Which wasn't his fault, but I think it is a really messed-up way to do a chemistry test. Imagine it happening at your job interview.

I did the same thing that I always did. I snapped out of my body and floated up to the ceiling.

I guess we had chemistry. I got the part of Amy Blue.

IT BEGINS

It doesn't happen that way, being discovered. An incredibly rare thing happened to me. That Gregg Araki and the investors took a chance on a complete unknown is like one in a million, maybe more. Interviewers have asked me what I'd have done if it weren't for acting, and my answer is usually, "I'd have been a pathologist or archaeologist," but it's probably not true. The reality is I don't know what I would have done, because frankly, it was always a foregone conclusion that I was bound for movies in some way. My life was always going to be big even when I tried to make it small.

Doom Generation was like acting boot camp for me. I knew very little about technical aspects of filmmaking. They would tell me: "Stand on your mark." I didn't know what a mark meant—I just wandered across the room. "The *X* on the floor. That's where you stand." They said: "Watch your camera framing." That means how much room you have, like in the invisible box around your head from the camera's viewpoint. You can't step too far to the left, or too far to the right, or you'll be out of the shot. "Let's rehearse blocking." Blocking is where you go from point A to Z in any given scene. You have to remember all that and then forget all of it to make it natural. It was a difficult shoot because we had a very short amount of filming time, and we had about three takes' worth of film for each scene. I tried my best to get everything right all the time.

My character, Amy Blue, was a snarling red lipsticked sixteen-year-old. I didn't know about creating a character, so I just based her on me at fourteen. A permanently pissed-off punk. I related immensely. Through Amy Blue I could channel my anger.

The first day of filming there was a huge earthquake in LA, the biggest in decades. I remember it was like surfing in bed, and I kept laughing, thinking it was super fun. Later I found out people died, so it was not so fun in reality. We filmed mostly at night in abandoned places on the edge of the city, for a postapocalyptic vibe. I thought the aesthetic was really cool. Gregg Araki told us that we'd be tired and cold, so I knew what to prepare for, and he was right.

One of the two male stars, very good looking as you'd expect, was permitted his bad behavior. It was the first time I'd heard about method acting, where the actor stays in character all the time, even when the camera isn't rolling. I have never met a female method actor. To me, "I'm a method actor" is usually synonymous with "I'm going to be a fucking dick to everybody on set." It's something so many young male Marlon Brando wannabe actors do. And I've never been on a set where that bad behavior wasn't indulged. I wasn't impressed, and I wasn't turned on by him, either, which I think may have confused him. He was incredibly mean to me. My boyfriend had just died. Everybody knew about Brett. I was shocked, but it was like nobody cared.

I quickly learned nobody was going to protect me.

There was a defining moment when the favoritism, the misogyny, the toxic environment on movie sets became real for me. But I didn't know how to articulate it then. I didn't understand why these guys were allowed to get away with everything and permitted all. We were shooting a scene with the actor and me in the front seat of a car. The director and cameraman were in the backseat, filming my close-up while the male actor gave me my lines from off camera. All of a sudden I felt something wet under my skirt, and an insistent pushing pressure on my vagina. The actor had taken a bottle of water under my skirt to spray and push onto my privates. I froze. Then I snapped. I went to lunge for him, but the camera was in the way. Gregg Araki just said, "Oh, children."

Later I told that story in an interview, and Gregg Araki wrote a long response that I skimmed and mostly don't remember except for the last part where he said, "Rose remembers it wrong." To me that is the height of misogyny and victim blaming. Gaslighting. Don't gaslight me, motherfucker. My vagina remembers. My body remembers. It is a scientific fact that memory changes every time we bring it up, *but* the body has memory that is even more accurate than the mind. Women know when they have been violated emotionally, physically, or verbally. And no man has the right to tell us otherwise. Our bodies shake, they burn, they do all kinds of things when they remember. Our muscles remember. We know by the way we feel when we have been violated. Even when we are drunk we know the difference between welcome and unwelcome—the body ALWAYS feels it during and after.

The male actor has since apologized and it is an apology I completely accept. There is no bad blood between him and me. I think he's great. I still respect Araki's work and am still grateful that they took a chance on me, but everyone on sets should be respected, not just the males.

The producer was a young woman at the time, although still the better part of a decade older than I was then. I ran into her years later and I brought it up with her, how disappointed I was that she didn't protect a young girl.

That happened over and over in films I did. It was like they didn't really even see me. I was just the girl and it was okay to be treated poorly. But it's not okay. So I did the only thing I could do. I pretended like I was one of the people in my books, the characters that I'd turn into when I read as a child. I created my own character in my own version of the movie that I had in my head. Later on I realized that the movie in my head was often a lot more interesting than the actual movie I was in. I eventually came to look at it like I was doing performance art in other people's worlds, like what Cindy Sherman did in photographs, transforming herself into other people and other lives. It was an attempt to preserve my sanity and the part I most enjoyed.

In *The Doom Generation* I had to shoot a three-way sex scene, which actually wound up being almost comical, because we were all wearing sweatpants and slippers and just making sexy noises. On film it looked super sexy, but in real life it felt as if there were eerie sex-cult parallels that I couldn't clearly articulate, even to myself. At the time I thought the language in the film was like 95 percent made-up slang. Some years later I did the DVD commentary, and I finally was adult enough to understand what I had been saying in the movie. Holy shit. No wonder my dad chased Gregg Araki out of the Seattle Film Festival and tried to beat him up.

In January 1995, I went to my first Sundance Film Festival with *The Doom Generation*. The excitement was palpable in the thin mountain air of Park City. Some people were so upset by the film that they walked out of our screenings. It was a very polarizing movie—nothing like a little nihilism to freak people out. I loved it. Despite my experiences on the set, I thought Araki and all of us involved in that film created something really cool and different. I would go on to be nominated for the Best Debut Performance at the Independent Spirit Awards (like the Oscars for independent films).

There were lawyers everywhere at Sundance. I didn't understand why they were all giving me their cards. It reminded me of my runaway days, with troll men creeping around trying to prey on you. I had no idea why I'd need a lawyer. I had come into the whole industry backward and because of that didn't know the "rules." Normally you come to Hollywood, enroll in an acting class, do workshops hoping an agent may choose you, start by trying to get on a commercial, work your way up to a one-line role on TV, maybe play a dead body or something, then try for a featured guest star, then a little role in an indie film, then hopefully a bigger one, and then the Holy Grail, starring in either a TV show or film. As you slowly climb the ladder—normally it takes years and lots of luck—you learn who the sharks are. But not me. I was a baby in the industry, and working at a level beyond my understanding. So I had never heard the rumors of what creeps to avoid, who to trust, if anyone. I was so innocent to the games played, it just didn't occur to me that people would lie to me, that they would have nefarious motives. I didn't know I needed to be scared of those who smiled and extended a hand to me.

There was one lawyer at Sundance, a little guy in cowboy boots, who approached me and explained why I needed a lawyer. He signed me and he got me an agent, and it went from there.

They immediately put me in a Pauly Shore movie because I needed money for rent. Sure, I had done my first lead in a film, which I think people assume means you're rich, but I wasn't paid enough money to be able to afford more than a tiny apartment, which I shared with this girl Julie, who was six foot one with short platinum-blond hair, a loud Australian who sounded like a giant toucan bird, squawking all the time. I loved her.

Being that I was now an actor, I kept waiting to feel like I had caught the "acting bug." I read about other actors and their passion for the craft, and I kept waiting for that passion to hit me. I loved film so much, I loved how sets worked and that we could make things that affected people and shaped minds, but the bug wasn't coming.

The indignities that went along with being an actress, I resented greatly. You know what it's like to go on a job interview and the stress/discomfort that goes along with it? Now, imagine that in your job interview you're required to break down and sob or laugh maniacally, all while looking sexy. Imagine being asked to do humiliating things like stand and turn so they can see your body, a group of men staring at your ass and tits but pretending it's about the role.

Auditioning for me was always traumatic. Like that hallway scene from when I was a child in the cult. Here's how it would go: I walk down a hallway, my heart is already racing, and I feel short of breath. There are maybe ten women on one side, ten women on the other, on little folding chairs under fluorescent lights. Everyone is in a teeny dress and heels or trying to look like the part, whatever it may be. I feel a surge of embarrassment being there with my hand out asking for the powers that be to choose me. I hate it. I hate every minute of it. The women give me the once-over and try to vibe me out. It feels like my skin is being pulled off with needles because I'm in a cattle call and I can see what I'm up against.

The night before, I worked hard to memorize the two scenes from a script that some mediocre asshole guy wrote and thought was great. Whatever, I need the job. I'm required to scream and wail for the part. I'm sitting outside in the hallway now with these ten to twenty other actresses and I can hear the muffled screams from inside the casting room. I try to tune the wailer out. After waiting an hour or so, it's my turn. They call me in. My heart starts racing even faster. My hands sweat, and my chest is turning red; I can feel the color splotches rising.

I walk into a small boxlike room; there's a camera set up to film me. There are five men in the room with a female casting director and the assistant operating the video camera. They're sitting in a semicircle staring at me. Hideous fluorescent lighting once again. I wonder how anyone ever gets hired under lights like these. I say hello to everybody in the tiny bit of time I have to impress them. My hands shake the script papers I'm holding. The reality is that behind the scenes they most likely already know who they're going to cast. They have offers out to stars while I'm in auditioning; they keep the auditioning going to have leverage and maybe a backup plan. I know that they're looking for a bigger name. I still go on the off chance that if something doesn't work out, maybe they'll pick me.

I hold my script scene papers in my shaking hand in case I forget the words. I proceed to do the scene. I, too, start wailing and crying at the appropriate moment. I wonder if everybody else out there is thinking the same thing as I'm thinking, but then I think maybe not because maybe they want to be here. Being that I was the accidental actor, I don't want to be here. I find it excruciating, very embarrassing, and I resent being treated like cattle, just one of many. I have tears rolling down my cheeks. They ask me to do the second scene again. Is that a good sign? "Thanks, Rose, good to see you." The casting director dismisses me. I shake her hand and say, "Thanks. Okay, bye." All that crying and screaming, I'm just going to shove back down inside of my body, my poor body that doesn't know what's going on or why I dredged up intense feelings.

Now I have to walk the gauntlet down the hallway past the other actresses staring at me. I seethe inside. I walk down the street looking for a trash bin to ditch my script pages. I always throw the script pages away as soon as possible because I don't want to be seen as an actress walking down the street; I don't want to add to my walk of shame. In two days I have another audition and I'll have to do it again.

I was sent in to audition for the role of Tatum Riley on an upcoming horror film, *Scream*. I was hoping they wouldn't lay me on top of anybody this time. Thankfully, it was just the usual screaming, sobbing bit, and I did well enough. I got a call from my agent and was offered $50,000 for the part. Holy cow! That was the most money I had ever heard of coming my way. Protocol would have been for my lawyer to counteroffer $100,000, and then I'd wind up getting $75,000, but my lawyer went back at $250,000. This so infuriated the head of the studio, he made me retest (a filmed audition) for the role three more times, even though I'd already had an offer. To me, it felt like the studio head wanted to humiliate me and penalize me for my lawyer's pissing contest move. That's what they do. Punish the girl for the actions of the representative.

Before I could retest, they hired Neve Campbell, an actress with dark hair. I thought, *Oh, God. They're never going to hire me now, because I have dark hair, too*. Those are the rules. I mentioned to the blond producer that I was thinking of going blond. Everyone's eyes lit up, *ding, ding, ding*, because you know, you can only have multiple females on-screen if they have different hair colors, because otherwise they think the audience is too stupid to tell them apart.

I had no desire to be blond, but I knew that was the only way I'd get hired. One of the producers promptly took me to her hairdresser who turned me into a midwestern blond. My plan worked. I was officially offered the part, but for less than all my counterparts because of said money-pissing contest. After paying my agent, manager, and lawyer, I wound up probably with $12,500. But it was still the most money I had ever seen.

My character died in *Scream*, but I wanted her to be more than a disposable young woman in a horror film. I was determined to make people feel for my character, Tatum Riley. I did not want her to go down without a fight. I wanted her to be memorable as a human.

No one discusses how disconnected people are when they watch horror films. God forbid you watch someone die awfully and you feel something. If you don't think this translates to real-life numbness, you're deluding yourself. Numbing yourself to violence against women comes early and if it's not coming to you at your home, and I hope it's not, then it's coming to you through TV and film. What I thought the original *Scream* did very well was to make you *care* about each character. We weren't disposable.

I'm proud of creating an indelible character who had one of the all-time most memorable deaths on-screen.

I am also proud of doing my own stunt work in the famous garage door death scene, the scene where Tatum dies. I ended up with bruises from my shoulders to my waist, but I knew that it would look better.

When I arrived on the set of *Scream*, who should I see? The same old creepy guy from my job as an extra on *Class of 1999*. The guy who molested me. He saw me and said, "I know you from somewhere. Don't I know you?" My heart started racing. But I had blond hair, so he didn't quite recognize me. He got replaced a week into filming, which made me incredibly happy.

Aside from that, the set of *Scream* was a refuge for me. Wes Craven was a special and complex man. He grew up speaking in tongues in a Baptist church. He came from Ohio, where I think he was a teacher, and he left with his young children or child and his wife, moved to New York City, became a taxi driver, and made his dreams come true. I totally respect that. He treated us actors like his equals, and it was a very special environment. Wes Craven was so kind, a true gentleman. I thought all my movies with big directors were going to be like this one. I was wrong.

We all kind of knew we were making something sort of magical, but we couldn't anticipate that it would become the phenomenon it did. Especially not me who knew nothing about box office numbers and things like that.

I got a puppy, a Boston terrier, Bug, a week before I went to film *Scream*. Bug was an extraordinarily trippy dog. She was on sale at the Beverly Center, a big ugly mall in Los Angeles. I walked by one day, and I saw this tiny black-and-white thing with two giant eyes going different directions, her paws stuck through the metal cage wires. I noticed her price, and that she had been put on sale twice, and I thought, *I understand what it's like to be discounted. This is the dog for me.*

Wes Craven fell in love with Bug, as did I. She grew up on sets. She understood that she had to be quiet during takes. She was flawless. She would be perfectly still and not even jangle her collar. She was photographed many times, with my boy dog, Fester, another Boston I got a couple of months later. They were photographed by Bruce Weber, Ellen von Unwerth, David LaChapelle, some of the biggest photographers ever. Bug was quite a catch on the set, let me tell you. She added that extra "oomph" of weirdness.

Shortly after *Scream* I went to the dentist. I was lying prone, with my mouth open as wide as it could go, in a fancy Beverly Hills dental office when the dentist stared into my mouth and declared: "You do not have movie star teeth." And I believed him, even though at this point, I was, technically, starring in movies.

As if I wanted some Hollywood bullshit teeth like his, with fake overly white Chiclets gnashing at the prospect of profiting off me? That was what my better self was saying, but how do you keep that better self intact when everything in the Hollywood system, in the media world, in the training you get as a young woman in this society tells you to do away with your realness?

The idea of not having movie star teeth stuck. I was sure when *Scream* came out, my crooked teeth were going to look five feet tall on-screen. At the premiere it was all I could focus on. My fucked-up teeth. I should've pointed out that I *was* a movie star even with my crooked teeth, but I didn't go that route. I started to straighten my teeth, something I actually regret doing. Most especially because it was an implanted idea and not one of my own. The bullshit brainwashing had begun in earnest.

The bigger picture is that I was the one whose face was going to be blown up and sent around the world, to influence the rest of the population to look homogenized. Little did I know that by being on-screen, I was a stand-in for all women. That was my role. I just didn't know it yet. And my straightened teeth were part of that messaging.

DEATH OF SELF

Most people have certain landmark experiences or events that become major milestones: high school, prom, college, wedding, that kind of stuff. Mine were truncated into an on-screen version. *Scream* was my version of college. And if *Scream* was college for me, *Jawbreaker*, which came next, was going backward into high school. It would be the only prom I would ever attend.

In fact, my character, Courtney Alice Shayne, doesn't just go to prom in an amazing gown, she wins prom queen. It's all going well until someone plays a recording of Courtney confessing to murder over the loudspeakers, and then the entire student body turns on her, pelting her with their corsages and cursing her.

So that was my prom. I loved the character I played. She was amoral, but I think sociopaths don't tend to be aware of the fact that they're sociopaths. So she just thought, *What's the big deal?*

I saw a classic film where the actress Gene Tierney played a sociopathic character. The film was called *Leave Her to Heaven*. In that, she pushes a little kid in a wheelchair off a cliff. When her husband says, "Why did you do that to Timmy?" she responds: "But darling, we needed more time alone together." I always thought that was bizarrely hilarious, and so I based Courtney Alice Shayne on her, a character that was my tip of the hat to classic Hollywood. With Darren Stein, the great writer and director of *Jawbreaker*, I turned Courtney Shayne into an iconic character. The other girls did a tremendous job in that movie as well.

In 1997 I had started doing another movie, *Phantoms*, made by the same studio that did *Scream*, Miramax. While still filming I was sent back to the Sundance Film Festival in late January. This time I was to be the belle of the ball. I had four movies at Sundance that year: one short and three films.

One of my films at Sundance was called *Going All the Way*. It's set in the 1950s, a beautiful little film, with a great production designer, great costume designer, and a great director. Jeremy Davies, the lead actor with whom I was friendly before we started shooting, was in it, too. I had to do a topless scene with him: in the scene I'm trying to turn him on and he can't get it up. I'd done topless scenes in *Doom Generation* and I thought this time it would be easier since I knew the actor previously, but it wound up being way more difficult. It was harder to detach from my body as I had done during *Doom Generation*, and I felt it this time. I cried after filming the scene.

At the Sundance premiere for *Going All the Way*, I took my seat in the theater, my heart racing with nerves as I would shortly be seeing myself on-screen, something that I couldn't get used to. My manager, Jill, who was seated next to me, whispered in my ear, telling me the head of the famed studio Miramax was sitting behind me in the theater. Miramax, owned by Disney, was the superpowerful company that owned the company that produced *Scream* and was producing the movie I was currently starring in with Ben Affleck, *Phantoms*. In fact, it was my second film with Affleck because he was also in *Going All the Way*. The lights dimmed in the theater. I had seen the Studio Head's name in the credits of *Scream* but had never seen his face. I didn't want to turn around and make it obvious that I was looking, so I kept my eyes forward. I went back to watching the movie. The topless scene came on and I wanted the ground to swallow me up. It had been really hard shooting that scene in the first place and I sat there remembering how I cried after filming it. I slid farther down in my seat. I noticed my manager turning and nodding in the direction of the Studio Head. When I replay the chain of events in my head, I'll always be chilled by that nod. I wondered what the nod meant for twenty years. Now I know.

By now we all know the Monster's name, but I have made a choice not to use it. I do not like the Monster's name, and though I know it, and maybe you know it, I refuse to have his name in my book.

When the film ended, the Studio Head was gone; he must've left early. Jill was superexcited when she told me that the Studio Head had summoned me to a business meeting the next day—a 10:00 a.m. meeting at the fanciest hotel in Park City, the Stein Eriksen. I was to meet him in the restaurant. Later on I would wonder whether the meeting got set up *while* we were watching the movie. Jill told me the Studio Head was a known star-maker, that this was my big chance to make a great impression. I told her since I already was in two movies produced by him, it was a safe bet I already had made a good impression and asked why I needed to go to the meeting if I was already employed. But she insisted the meeting was necessary and that I go. I said yes and the meeting was added to my already packed schedule the next day. Jill told me that this man wielded an incredible amount of "power" in Hollywood; I could sense it by the way she was breathlessly talking about him. I was so new to the industry's upper echelon, I didn't know anything about this man or what exactly his kind of power meant. I didn't know what so many already knew, that he was a predator and I was walking into a trap—a trap that started the twenty-year conspiracy of collusion.

The next morning at Sundance, I got up early, ready for my meeting with him, before the full day of press I had to do for the other three films I had at Sundance. I was overwhelmed just thinking about my to-do list, but I figured with Jill the manager at my side, I'd be able to pull it off.

I have thought a lot about the day my life got hijacked by evil. There was an MTV camera crew following me around that day. "A Day in the Life of Rose McGowan" was the theme. The MTV crew was to wait outside the front of the hotel until I returned from my meeting. I've kicked myself through the years because before I went in to this fateful "meeting," I turned to the MTV camera and, with a big genuine smile, said, "I think my life is finally getting easier."

I thought since I was now achieving a little financial stability and was in nicer environments, that it meant life was easing up a bit. I was tired of surviving, I was tired of fear, I was tired of pain, I was tired of hurting. I wanted to soar. I wanted to fly. I wanted to be free. Instead I got a prison sentence.

I waved good-bye to the camera, walked into the hotel, and made my way to the restaurant. I said a chipper "Good morning!" to the grim-faced restaurant host. He told me that the Studio Head was stuck on a call, working from the office in his suite, and that I was meant to go up and wait. I smiled at the host and thanked him, but he turned away and didn't smile back. I remember thinking, *Well, he's not very friendly*.

I found the room number and knocked on the door. Two male assistants opened the door. I offered them another chipper "Good morning!" The men said nothing and looked down. I thought, *Well, gosh. They're not so friendly either*. They pushed past me silently and left me alone to walk in the room.

When most people hear "hotel room," they imagine a bed, dresser, and little bathroom. The Monster's hotel room was the entire floor of the hotel. Probably about three thousand square feet, like an average- to large-size house, definitely not the small hotel room the mind conjures.

I went inside, and into the largest living room I had ever seen, and there my boss was, the Studio Head himself, on one end of a huge couch, talking loudly on the phone. He gestured for me to sit down. He kept talking loudly on the phone while I waited for about five minutes for him to finish his call. I had time to study him. I was repulsed immediately.

The Studio Head was not attractive in the least; let's just say it would be the understatement of the century to say that he would never win a beauty pageant. Well, maybe a beauty pageant in hell. He might win the award for most monstrous. He's a very large person, vertically and horizontally, oily skinned, his face pockmarked, with a bulbous fleshy nose and liver lips. His right eye squinches up more than the left, giving him a lopsided look. He reminded me of a melted pineapple.

He could be the bogeyman of your worst nightmares.

He became mine.

I sat on a far end of a couch and he sat about five feet away while he talked loudly on the phone. I stared at the ceiling so I didn't have to look at his face. Eventually he finished up his phone call and we proceeded to have our meeting. I figured I'd do what I did best, use my intelligence and wit to prove I was different from the stereotype of an actress. Even being new in my career, I already resented being lumped in with the actress cliché. I was certain we would be working together for many years to come, and we were here to plot out the grand arc of my career. That's how my manager had framed this very significant meeting. Those were my earnest, enthused expectations. I mean, if you were in the middle of your second extremely high-profile job for a company, would you not assume that you were already a valuable employee with a big future? Looking back, I want to hug myself for being so naive. I really thought that this was a business meeting and that this creep cared what I had to say. I was so wrong.

Later I'd find out that other actresses had been warned about what could happen when this Studio Head summoned you to a meeting. Later I'd find out lots of things. Completely unbeknownst to me, an industry newbie, this hideous beast had a long track record of preying on young women. Turns out even in 1997, this was an open secret in the industry. Sadly, it seemed everybody in Hollywood on the business side knew that if you got summoned to a meeting, it was probably going to go differently than you expected.

But *I* didn't know the rumblings and the secrets, the gossip and the warning signs. As a street kid I had known to be on guard for the trolls. It just never occurred to me that the head troll would be in Hollywood. That these fancy people were in fact more dangerous than those on the streets. These people in their $3,500 suits could be far more evil than a guy who's just cruising you to get ass one night on the street corner. You know why? Because these are the creeps who spread their rot to our world.

The Studio Head asked about the kinds of projects I wanted to do. *What a banal question,* I thought. I said I wanted to do projects that mattered. He seemed like a smartish man, gruff, rough around the edges, not particularly well-mannered. Piggy, to be honest. I was taken aback by his size and ugliness. I covered it by talking about my love of classic films. Apropos of nothing, he said he had a Jacuzzi in his hotel room. Okaaay, non sequitur much? I figured the Jacuzzi comment was a brag about being rich and having a Jacuzzi, which I thought was tacky. I didn't realize it was part of the setup. I didn't know what to answer, so I continued with my story. At about 10:30 a.m. we wrapped up the meeting and he said he'd walk me out. I thought, *Well, that went well.* It was my first meeting with a bigwig. I couldn't wait to tell my manager, Jill, all about it. I figured she'd be proud of me.

I exited the living room and walked down a hallway. Walking behind me, his huge size seemed even more overwhelming. I'm five foot four. It seemed like he was six feet, four inches tall and six feet, four inches wide. He probably weighed three times as much as I did, if not more.

My brain was already outside and onto its next task, the exit interview. The MTV camera was waiting for me outside and would be filming as I exited the building. I was hoping my makeup still looked good. In the hallway, we passed a door. Suddenly he stops me and says, "This is the Jacuzzi room." I didn't know what to do, so I politely looked in and told him it looked nice. I had no idea why I was being stopped to look at a Jacuzzi when we all know what they look like. I made the appropriate *you have a nice Jacuzzi* sounds. It's superhot and humid.

I feel a hand on my back and it pushes me farther into the tiny, dark, incredibly hot room. Everything at this point happens so quickly, and yet so slowly. I'm confused as to why I'm in this room. I can't breathe. He is standing right in front of me, taking all the space in the wood-paneled room. It all happens so fast. My clothes are getting peeled off me. I back into the wall, but there's nowhere to go. I freeze, like a statue. I don't know what's happening; my sweater is being pulled over my head and his hand pulls my pants down. He bends over and pulls my shoes off. I am now naked. This all happened in the space of about thirty seconds, it feels like. My brain is trying to play catch-up. Alarms are ringing loudly in my head. He takes his clothes off. What the fuck is happening? I'm picked up and placed on the edge of the Jacuzzi. I am naked, up to my knees in the hot water. I curl into myself. I did what so many who experience trauma do, I disassociated and left my body. I went up above myself. He gets in the water with a large splash. He pushes me up against the wall. My knees are pressed together. He places his hands on them and pushes my legs apart. I am open wide to a monster. Literally more naked than I have ever been. He places his monster face between my legs. Alarms keep blaring in my head, *Wake up, Rose; wake up,*

Rose. But I was frozen like a statue if the statue's legs had been spread wide.

Anybody who's a sexual assault victim will tell you: the trauma does strange things to your sense of time, your memory. There are details you remember with uncanny accuracy—the shape of the tiles, the yellow quality of the light, the obscene bulbousness of a nose out of all proportion to the rest of a face—and then there are gaps in the timeline where there's nothing, nothing. Every second extends for a hellish eternity, but it all happens in a flash. And your life is never the same. My life was never going to be the same.

Detached from my body, I hover up under the ceiling, watching myself sitting on the edge of the tub, against a wall, held in place by the Monster whose face is between my legs, trapped by a beast. In this tiny room with this huge man, my mind is blank. *Wake up, Rose; get out of here.*

I try to make sense of what the fuck is going on. How did I get pulled into this position and pushed up against this wall? When did my clothes come off? I don't know what to do. It goes on and on and on. My skin feels like it wants to fall off. His disgusting tongue is INSIDE of me. Oh my God. Tears roll down my face. The water is splashing because he is grabbing himself underwater. One hand holding me, his other holding it. His tongue stabs inside of me again.

Wake up, Rose. My brain starts to scramble. Survival instinct kicks in and I am desperately trying to figure out how to get the fuck away and make it stop. He is slurping and smacking his disgusting lips on me. His fat tongue wet all over my most private parts. *Oh my God, stop.* I don't know how else to get out of this situation, so I remember the *When Harry Met Sally* movie with its big fake orgasm scene. So I did that. I pretended to have an orgasm. I moan loudly, over and over and over, tears falling down my face, mingling with the sweat of the room. He moans loudly; through my tears I see his semen floating on top of the bubbles.

The fake orgasm works. He seems satisfied, and he sets me down; my legs feel like jelly. He tells me to get dressed. I grab a towel and hurriedly dry myself as best I can. My whole body is shaking.

I try to find my clothes. I'm in total shock and moving somewhat mechanically. I'm still hovering up above, not quite in my body, and I'm trying to put my clothes on and make sense of what has just happened. It's like a race you can't keep up with. My life has been rerouted. I just got hijacked.

Later, as I replay what happened over and over, I flash back to those men I saw on my way to that fateful meeting. The grim-faced restaurant host, the assistants who wouldn't look at me: they did this to me just as much as he did. And I hate them for it.

I stumble out of the hotel in a state of shock. The MTV camera crew is out there filming, the camera rolling. The first thing I see when I come out is them with their microphone held out in my face asking me how it went. That footage exists somewhere. I hope I never see it. I immediately get taken to a photo op with my costar for *Phantoms*. I am shaking and my eyes fill with tears; I say where I've just come from, and my costar says, "Goddamn it. I told him to stop doing that."

I don't remember much else from that day other than getting a plane ticket and going home. I wanted to go home and see my friend Ingrid who had also been sexually assaulted. I knew I could talk to her. She was my best friend.

The Monster heard I left town and kept trying to call me. He left me messages telling me I was his new special friend. He named other big actresses who worked with him, who won Oscars, and said they were his special friends, too.

I threw up when I heard his voice. There was no way I was returning his calls.

I felt so dirty. I had been so violated and I was sad to the core of my being. I kept thinking about how he'd been sitting behind me in the theater the night before it happened. Which made it—not my responsibility, exactly, but—like I had had a hand in tempting him. Which made it even sicker and made me feel dirtier. I know other victims feel this way too. We replay the tape of the event over and over, blaming ourselves. If only, if only, if only.

I thought of how I had turned to the MTV cameraman and said I thought my life was finally getting easier. I wondered if in that moment I'd cursed myself, I'd jinxed it, even though I know logically none of this is my fault.

It was criminal on every level: he was my employer, I was his employee; he was a tremendously influential and powerful player in my industry, and I was a newcomer just barely making a living. He was a huge ogre of a man, and I was a girl. I am crying while I write this.

"I told him to stop doing that." That comment has haunted me. How fucked up that everyone shook their heads and just looked the other way. But the cover-up was just beginning.

During the immediate period afterward, I couldn't stop crying. One of my calls was to my manager. It was so fucked up, she counseled me to see it as something that would help my career in the long run. I threw up. I felt like I was in a fun house and all the mirrors were reflecting my horrors. And my manager's instinct was to squash everything, which just freaked me out more. How could she not have known? And if she did, how could the woman I trusted with my life set me up? I was terrified. I had fallen into a backward, fucked-up world.

I called my management agency. The man who answered was a player, a powerful guy in town at the time. I told him what happened to me. And he said: "Goddamn it, I just had an exposé about him killed in the *LA Times*; he owes it to me not to do this."

Oh my God. This man could have stopped this Monster from hurting me, but instead chose to do him a solid. It's okay, right? I was just a girl. My brain was stunned into silence. Who were these awful people?

I wanted to press charges. Someone connected me with a brusque female criminal attorney who said, "You are an actress. You've done a sex scene. You'll never win. You're done." I went cold all over. I was alone. I was all alone.

I knew if I came out publicly with this, nothing was going to happen to the Monster, but I—I would never work again. If I lost work, I wouldn't be able to support myself, and once again, I was terrified of being homeless. No work would land me back on the streets, and homelessness was a death sentence. I knew if I died I'd be remembered for revealing my rapist, but not for my achievements. I didn't want his name next to mine in my obituary.

I thought a lot about death during this time. Mine. His. And everything felt dirty in my world.

I got a call from the head of my then law firm. I had to tell him the truth once again, feeling violated by sharing a deep open wound with some industry creep I didn't even know, which felt violating in and of itself. The big lawyer urged, "I really want you to publicly come out against the studio head. It would be a great thing to do." But it wouldn't be a great thing to do. As battered as my spirit was, I instinctively knew that I would become a minor player in some kind of power game between two powerful men. Thanks for nothing.

Even in my traumatized state, I realized that even this powerful man couldn't do anything that would help me. Maybe if there was enough of an incentive, a financial or political win, they might consider breaching the "honor code" that protects these fuckers. But probably not. And as it turned out, no one did. Not for twenty years.

Because it's okay. It's the business. And it is just. a. girl.

So I knew better than to say anything. I was not going to be used as a pawn by these people. Even in my messed-up state, I refused to be a pawn. I didn't realize yet that I already was one.

I was so disgusted. I'd worked so hard my whole life trying to survive. I'd already been groped, grabbed at, hollered at, diminished, fetishized, but this was a whole other level of violation.

Traumatically, I had to go back to work and finish *Phantoms* as I was in the middle of filming when I fatefully went to Sundance. I played a sixteen-year-old in the film, but now I felt like I was around a hundred years old. I was so disgusted with Hollywood, but I was under contract and I had to finish. I had to hear his pig name every day, over and over and over.

So what could I, a young powerless woman, do? I wanted to put him on notice that I was not okay with what he did. I was pretty broke still. I said to my attorney, "I'm going to need money for intensive therapy. And I'm going to need money to donate to a rape crisis center." My lawyer got me a hundred thousand dollars. That money felt dirty, anyway. I largely gave it away. It brought me no solace. But it was the only way I could put the pig on notice that I was not okay at all with what he did.

I started to hear rumblings around town. Snippets here and there. The Monster was blacklisting me. I heard he called every other studio and independent producer in town and said, "Don't hire her. She's bad news."

So many people heard about what had happened. It had spread like wildfire through Hollywood. One assistant tells another assistant, one producer tells another producer, and on and on. It seemed like every creep in Hollywood knew about my most vulnerable and violated moment. And I was the one who was punished for it. It's like being assaulted over and over and over.

People think that you can get over being assaulted. The thing with trauma and rape and sexual assault is that it freezes in your mind as if it happened yesterday. It's very, very hard to get rid of, because a large part of you, the you that was whole, has been murdered. I've come to a certain peace with it, but my life will always be irrevocably tied to this Monster because of what he stole from me. Because his desire to dominate superseded my right to bodily integrity, my right to be whole. Sexual assault takes away our ability to be who we were and steals who we were meant to be. Now we victims are cast in a role we didn't want to play. Girls grow up being terrified of rape because it's allowed to happen. Girls are told in health class, just like I was, that it's best to submit and be pliant, that way you can live. Yeah, my body might be alive, but who I was is dead. I'm now a live body carrying a deadened spirit around. And it's allowed to go unpunished. Everyone just wants it to go away so *they* can feel better. But what about us? How do we feel better? Who cares for us? Don't you dare teach us girls to be pliant; teach your boys not to rape. To me rape cannot be defined by a law a man wrote. How would that man know what rape is? Rape to me is any violation of my body. If you enter my body via tongue, fingers, penis, object without my consent, that to me is rape and I need no law telling me what I know to be true.

I wanted to go back to before when I was a whole person. I wanted to go back to being a strong badass, but I was now in a million pieces. I couldn't stop crying. I couldn't stop the screaming nightmares. I couldn't stop Hollywood. I just wanted out and away. My body kept having its own flashbacks.

THE PREMIERE

After two hours of being turned into a plastic doll-like version of myself, I get driven in a very cold air-conditioned car to the red carpet. As I get closer, I can hear the people on the sidewalk screaming. My stress level is going through the roof. My hands shake, my legs shake. I know I'm going to participate in a massacre of sorts by opening myself up to being bullied on a global level. That's really what it translates to. It's not so much, "Oh, let's go celebrate the opening of this project that I worked on," but more about getting savaged by trolls online for daring to exist.

I know that whatever photos are taken of me today will be on stupid gossip sites and magazines. They're going to say nasty things about me on all the message boards. I know this is going to happen, but here I am anyway. Getting ready to pose. I feel nauseated. I hate this.

I step out of the car and smile, a ridiculous expression on my face that in photographs reads strangely; sometimes I'd even laugh out of sheer hysteria. I have to wait with my publicist to signal me for my turn to walk the gauntlet, aka the red carpet. The celebrities having their photos taken are staggered so you're not in front of the cameras at the same time. I'm shaking harder now because of the yelling and noise, and I can feel my knee twitching and bouncing underneath my uncomfortable gown. It's now my turn. In little mincing steps—my stupid high heels are already killing my feet, my fake eyelashes are heavy every time I blink, and with voluminous hair that's puffed out to God knows where—I make my way to the first photographer. It's usually anywhere from ten to a hundred photographers, mostly men, screaming, raging, yelling your name as loud as they can to get attention, my body absorbing their yelling as aggression. Mince mince mince, three steps down, stop, hand on hip, smile, flash, repeat. The photographers yelling "Over the shoulder, over the shoulder!" because that way they can get both my ass and my face in the shot. I do what I'm asked to do and turn. There, now you have my ass, too; I've done my "job." This part of my job entails being a piece of meat to be consumed and savaged and judged. Fun times.

Now I've successfully made it to the end of the photographers' line and I move on to TV press like Entertainment Tonight *or* Extra, *shows that glorify banality. Their cameras pan up and down my body. They ask me to turn around, and once again, like the good brainwashed girl I am, I oblige. I do a slow twirl, feeling ridiculous, not knowing I could've said no.*

Hollywood thinks this is normal—they started it—but it's not normal; they've just spread this diseased idea of beauty to the world. And I am a part of that disease.

On the big screen, I see my name, I see my face moving, and all I can concentrate on is my teeth. My small nose looks giant. I didn't know I had that one wrinkle. I didn't know that my voice sounded so much like my sister's. Oh shit, here comes the topless scene. I feel a strange shame wash over me. Not because I'm naked, but because of how I'm interpreted by the men doing the interpreting.

CIRCUS LIFE

I was so sick of working. I'd worked to survive since I was a child, at any kind of job I could possibly get, and I'd never had any period where I didn't have to worry, like stomach-in-knots worry. When I was little, I heard kids in school talk about going on vacation. Once I figured out what they meant by vacation, I started thinking about what my dream vacation would look like. I decided mine would be a week in a hospital where no one could hurt me and I could have all the pudding I wanted. And right about then it was all I wanted. Pudding and no one hurting me.

My angel dogs, Bug and Fester, were my lifesavers during this time. They were what got me up in the morning. I had to tend to them. I couldn't wallow in my sadness. I hated how I was feeling after the assault. There was some good news around the corner, though, quite literally.

In the year leading up to our meeting, a huge rock star had a meteoric rise and his name was Marilyn Manson. I was quite unaware of what was going on in popular culture because I'd been so deeply absorbed into the world of movies. As far as Marilyn Manson went, I remember seeing a couple photos of him in makeup and thinking, *He is the ugliest person I've ever seen*. But simultaneously I thought it was kind of cool that he was willing to be that ugly.

Then one night I was at this absurd restaurant in New York, an S&M restaurant, of all ridiculous things, where you had to choose what kind of abuse you wanted while you ate. The whole place was dark so you didn't even know what you were eating, which, as a picky Virgo, was not my thing. Ceding control in exchange for food seemed silly and I was trying not to laugh out loud. Some people were in cages, and the waiter said I could go in, too. I settled on letting him rub my feet.

The waiter rubbing my feet said to me, "My friend Brian has a crush on you." "Oh, that's nice." He went on, "I'm from Ohio. We knew each other back home." "Oh, okay, uh-huh, that's nice." He said, "It's Marilyn Manson." And I said, "Oh, that's the ugly guy." He was taken aback. He said, "Yes, but he has a huge crush on you." I said, "Okay, well, that's nice."

Then one night shortly after that and back in LA, I was on my way to a screening of the seminal indie movie *Gummo*. I was late, and I had to bang on the screening room doors to be let in. I knocked again and finally this guy who looked like a cross between an eighteenth-century dandy and Ichabod Crane opened it. I looked right at him, recognized him as Marilyn Manson. I smiled and said, "Hey, I hear you have a crush on me." I didn't think he was ugly in person, I thought he was incredibly unique looking. We were pretty much together from that moment on.

For the record, you call him Manson. Anybody who would later come up and say, "How's Marilyn?," I would know immediately they didn't know him. Or they would say, "How's Brian?," trying to make it seem like they knew who he was, because they knew his real name.

We had so much fun. We truly did. But still, I was plagued with night terrors and post-traumatic stress disorder. The first year we were together, Manson patched me back together after the assault. I didn't tell him what had happened for the first few months, but finally he asked a girlfriend of mine, "What's wrong with Rose?" I was waking up screaming at night, soaking the sheets with my night sweats. My friend told him the truth, and Manson was so sweet with me. Finally, some kindness.

He was a very misunderstood person. Even though the media had dealt with controversial musicians like Alice Cooper or Ozzy Osbourne, and rational people knew it was all about art performance, with Manson they really bought it. They really thought he spent his nights skinning puppy dogs alive and boiling them in vats of acid while saying, "Hail Satan!"

In reality, the exact opposite was going on. The truth was that at the time when he wasn't creating electrifying music, Manson was painting watercolors of my Boston terriers while I was ordering glassware from Martha Stewart's online store.

We basically hid out from the world at home, totally domestic, when we weren't on the road having mad escapades. I was happy because I could forget about what happened to me, at least during the day.

Nevertheless, people thought it was bizarre that I was going out with him. When they freaked out about it, I'd think, *But you're bizarre to me. This is somebody who's kind to me, who's taking care of me.* Manson always saw to my needs and paid great attention to detail, and we fell in love. When I ran off with the Manson circus, I didn't really work for about three and a half years. We had to worry about death threats and bomb threats, and being terrorized online, but at least I didn't have to worry about where my next meal was coming from because I'd saved up some money from acting.

It was a blast, and we were madly in love, and anybody else who thinks differently is wrong. It was a pretty legendary relationship, not just in the media. It was a pretty legendary relationship behind the scenes, too. We had a whole lot of amazing.

Unfortunately, it seems his manager, his bandmates, and his whiny friend Billy were telling Manson things to the effect of "Rose is making you look soft. Going out with an actress is making you look like a pussy." Since when is the long tradition of actresses and rock stars going out together a bad thing? Whatever. Male jealousy is a strange and stupid thing. Manson likes yes-people around him. I've never been a yes-person.

I knew there were going to be serious career repercussions for me going out with Manson. People already didn't know what to do with me because I was unique, because I wasn't the "girl next door." I was told that a lot by casting directors, "You're not the girl next door." You know who was? A blond Reese Witherspoon type. So apparently everyone who has a female neighbor better make sure she looks like a sweet blonde who never offends. The first five months I went out with Manson I kept it very much on the down low because I didn't want to be known as his accessory. I eventually got brave enough and decided: you love who you love, this is it. I braced myself and came out and took the slings and arrows from the world. Boy, did I take some hits. The squares were after me, mainstream media in a tizzy.

Sometimes messing with the public was fun. I love subversion. I have an impish sense of humor. Hollywood is filled with nervous little biddies in the shape of men and women. They're such scaredy-cats. They're scared of their own shadows. They're scared of anything different, anything unique. Manson terrified them. When I'd see how these producers reacted when they saw Manson, I'd think to myself, *You're in the* entertainment *business, what part of entertainment do you not get?*

A few years ago, long after we split up, I was at a dinner with ten male agents from the "powerhouse" Hollywood agency CAA. If there's one thing I really loathe, it's these guys. The agents are such douche bags and have incredibly delusional senses of self-worth. They make the lists for who gets hired and pushed and who doesn't. One of the agents (think an older frat dude in a $3k suit) turned to me and asked one of the questions that most annoys me:

"Why'd you go out with that freak Marilyn?"

Before he could even get his offensive question out, I stood up. I walked to the head of the table. I put my hands on it. I leaned forward. I looked each one in the eye, and I said, "Gentlemen, let me give you some truth. The second you can make a child feel less alone, the second you can move somebody to tears, the second you can make somebody feel, think, and live through you by being a creative, by being an artist, the second you can do that, you can discuss why I went out with somebody who is more creative in his little pinkie than you will ever be in your goddamn boring life with your fucked-up value system and your bloated ego that is not warranted for any fucking reason that I can discern. Alone in the dark at the end of your life, you will be an empty suit in a coffin. You will leave nothing and be nothing. Good night, gentlemen." I stand by my words.

Even if Hollywood was scared of Manson, he obviously deeply affected many people. One time, I remember standing on the side of the stage, looking out over 350,000 people, this human wave of an audience, while Manson sang a song called "Coma White," which was a song about my life, my story.

There's something cold and blank behind her smile.

Yes, there was something definitely cold and blank behind my smile.

She's standing on an overpass in her miracle mile.

I lived in the Miracle Mile section of LA when we first met.

He goes on to sing in that particular song about taking pills to make yourself numb and dumb, but how all the drugs in the world wouldn't save you from yourself. At the time, I had started taking medication for my overwhelming depression and panic attacks after the assault.

It was a uniquely strange sensation to hear all these people singing along to lyrics that were about my life story, even though they didn't know it.

Touring was both fun and banal, and it was especially hard encountering groupies subjugating themselves and letting themselves be used and abused by the guys in the band and the crew, just for the sake of being closer to someone famous. At every stop of the tour I was on, women were lining up to submit themselves to abuse. Wash, rinse, repeat. One of the guys in the band would zero in on overweight girls, spend all night with them, and propose marriage. Then he'd never call them again, and do it the next night. I'd bet these girls have histories of personal abuse, of being valued only for one thing, and they think that by being with someone famous, or having a rock star pay them any attention, their lives are going to change. In the face of that, to keep my pro-woman stance alive was a bit of a challenge, but I did. I didn't know enough yet to know that the system is flawed because of men.

It was also lonely. Manson and I would be holed up in the back of the tour bus watching *The Big Lebowski* for the sixtieth time, while the other guys would be in the front of the bus or hotel doing whatever it was they did. Manson was quite shy at the time, and the guys were nice to me, but grudgingly and only if Manson was around. He didn't mingle much in the public, so I wasn't in danger of groupie girls coming after him. But a lot of the fangirls online hated me because I occupied a space they imagined themselves filling. It inspired a lot of rage toward me, just for existing. On the Hollywood side I had these idiots thinking I boil cats, then on the public side, these girls hating me for being with someone they fantasized about. Fun times.

I used to think, *What's the upside in all of this?* Well, the upside was that we stuck together. We were like a unit. And he did put on an incredible show. I loved dancing on the side of the stage because the band could rock the house down.

Once, on my birthday, Manson took me to Italy, to the Tuscan countryside where I was born and raised. I was trying to find the stone barn I was born in on the duke's property. There's Manson, all 138 pounds, six feet three inches, dressed in black, wearing big stomper shoes that made him six seven, looking like a scarecrow with his hat, and a leather jacket, in the heat of Tuscany, walking over hillsides with me, looking like a deranged Easter egg in a long pink skirt with a yellow shirt, and Petey, his gigantic bodyguard, following close behind. We huffed and puffed all over the hills, trying to find the barn. At one point this little kid on a tiny bike passed us and yelled out in his thick accent, "Ai! Ai, Marilyn Manson!" It was surreal and hilarious.

Eventually, we found the stone barn. The sister of the duke of Zoagli now owned the property—Rosa Arianna, my namesake. When I presented myself on her doorstep, I worked up the courage to knock. She came out and started screaming at me in Italian to get off her property, trying to hit me with a broom. I laughed it off as we scurried away.

But we found the barn I was born in. I was so touched by him taking me there.

The first big event I went to with Manson was a little thing in 1998 called the MTV Video Music Awards. At that time, everyone tuned in to see what people were wearing and what the incredible moment would be, as well as what the acts were going to be doing onstage, who was going to say what, and what crazy music-world stuff might happen. It was a really special time. Courtney Love and Hole were huge. Manson was the most controversial star in the world. There was some cool art going on in the mainstream.

Thinking about my own red-carpet moment, I figured, *Manson's going to be really flamboyant for these awards. What the hell do I wear?* And then I started to think about what I was going to wear in regard to how Hollywood and mainstream media would perceive me; I thought: *You know what? Fuck you. You want to objectify me? You want to see a body? This is what you want? All you media men, all you photographers, you vultures, this is what you want to see? I'll fucking show you a body.* And so I did.

Wearing the "naked dress," as I call it, was a big middle finger to pretty much everybody.

It was a reclamation of my own body after my assault. I wanted to challenge the media to see how they would deal with it. You want me to be your show pony, I'll be your show pony.

The infamous dress was sent to me the night before the event. The day of the MTV Awards I had a 101-degree fever and was taking sinus medicine that made me loopy.

On the way there, I had to be on my knees in the limo because otherwise the beading from the dress would have imprinted a waffle print on my backside, which would be more or less entirely exposed to the cameras. My heart was racing as we stepped out of the limo. I raised my arm up high and silently said, "Yes, bitches, I'm here." I changed right when I got off the red carpet into a different outfit, one that covered more.

The dress raised hell, which, I guess, was my intention of sorts, but what I didn't anticipate was the global slut-shaming that came afterward. I didn't realize how seriously everyone would take it. Do these people not understand rebellion and humor? I thought, *It's a music awards show. My dress was punk as fuck. This is what it's about.*

It was, of course, misinterpreted and sexualized, which was the exact opposite point I was trying to make. That's the thing that other women who have copied me have gotten wrong through the years; when they copy the dress, they do it to be sexy and turn society on. I didn't do it to be sexy. I did it with power, not to titillate or turn on the boys and men of the world. I did it as a big middle finger, and there's the difference.

I think that's why it became such an iconic thing. Every year, every time there's an awards show, that picture of me pops back up. It was my first time having a big scandal, so to speak, a global one, anyway. I say first time because I've done it several times since, but that one is the most physically memorable to people. For years, I was the actress who wore the dress. Regardless, it took some serious bravery to do that. I was scared, but I just did it anyway. Punk as fuck.

At least I know when I'm eighty, looking back on my history, I certainly won't identify as a scared person who didn't live. I'm from Europe, I'm not a freak about my body. American puritanical society shames you for daring to show any part of yourself, *especially* when it's done in a nonsexual manner. When a woman owns her body, she gets vilified. I was vilified for making people uncomfortable. Ever since then I've had to deal with the slut-shaming that came from wearing that dress. I regretted it at times, and other times I just thought, *Fuck off*. Now, I don't regret it at all. It's my body: I can do what I want with it. But being a part of a media shitstorm is definitely a weird version of Mr. Toad's Wild Ride and there's nothing that can prepare you for it.

At the time, all the managers and agents were saying to both Manson and to me, "You have to have a website about yourself and look at the message boards to see what the fans want." That was before famous people understood it's treacherous, and you absolutely should NOT look at message boards. It's like *Lord of the Flies*. They pile on the defenseless and beat them with their negativity. Some of the vilest scum of humanity post on message boards, bad people with bad energy. I guess now they're on all platforms. I can't imagine what these people are like in their real lives or why they have so much hate in their hearts. If you are one of them, know that you cause damage to other humans.

So I dove into the message boards and went deep. I took a lot of things into my brain that would break most people. Most people can't handle hearing one bad thing said about them, let alone global bullying. About your face, your body, your character. The enmity and the hatred for me was so intense. When you read nasty things about yourself, it sticks. You can have thirty compliments and one negative, and you remember the negative. It's human nature, right? Now imagine that times millions. It really messed with my head.

Just know that if people are saying cruel things to you and about you, you can weather it, you will survive. If you can get to a place where you are okay marching to your own beat, you will thrive and be free. I promise you, it gets better.

It took me time to learn this. One day, a supposed friend told me to check out this site online, so I did. It was a website devoted to how fat I was. I was shocked. It was page after page of me. In some pages my dog's face was mixed with mine; in other ones I was put on a cross with diagrams all over me pointing out how fat I was. It was really expertly done, and the crazy woman—yes, it was a woman who created it—could spell and used perfect grammar, so it really stuck in my head. Somehow it's easier to dismiss the people who write mean things when they can't even spell. Like "your a whore" instead of "you're a whore." When they can spell, it stings more. It was strange because the creator of this lovely site used a lot of the same terminology that I had used in my own head when I was at the height of my anorexia.

I became quite thin again after that. It stuck in my brain that I was fat. It reactivated my eating disorder. It fucked me up. Constantly "hearing" about how ugly you are, and then seeing that every photo of you has been retouched or photoshopped, you lose sight completely of what you look like. I just thought at one point, *I must look like Quasimodo.*

It went beyond skin deep. I was starting to lose all sense of who I was and what I was.

The trolling continued. When a loser named Perez Hilton started putting me on his (at the time) hugely popular website, he was merciless. He'd draw me with penises coming out of my mouth and semen all over my face. He said horrible things about everyone, but particularly women. He's definitely a woman hater. He would call Jennifer Aniston "Maniston" and think that he was deep with wit. Wow, sir, you are the next Oscar Wilde. Look at you go.

I always wonder if these dirtbags, after they're done for the day tearing other humans down, do they go home and go,

"Woo, that was a hard day at work drawing dicks on someone's face; I feel really good about myself"?

There's a saying, a loose statistic, that one fan letter equals five thousand actual fans, because only one out of that many is going to take the trouble to actually sit down and write a fan letter. So then when there are one thousand vicious comments, you really start wondering about society. To the creeps out there in the public, the ones who take the time to write these nasty things: rest assured knowing that you contribute to the breaking down of a person. Congratulations. Go, you. Your contribution to the world is duly noted. Reverse course.

More recently, I was on NPR, and I was curious whether the comments on their website would be more highbrow, more articulate. Nope. It was mostly a bunch of white dudes freaking out because I called them on their male privilege. It is so obvious, it's amazing when they refuse to acknowledge it. It is indefensible and true. And even on NPR's website, the trolls incorrectly taunt "your a whore."

People say that well-known people are worshipped, but I don't really see that. I see the horrible tearing down. You take these hits just for daring to exist on a larger stage than others. But why is that my fault? I'm not asking for deification. I'm just asking for simple respect for being someone whose only crime is being visible. All we are doing is being of service to you. There is also this conflated belief that if someone is perceived to have money, that person must have no problems and therefore it's okay to tear the person down. Almost all well-known people, minus those who are there through nepotism, come from humble backgrounds and we worked our asses off to get to where we are. So back the fuck up.

Fame is like you've suddenly moved into the tiniest town in the world with all the small-town gossips doing their best to shame the local girl they deem "loose." Just like in tiny towns, the gossips sit in their homes and peer out their curtains and track your every move, flaunting their precious bits of intel, exchanging it with others like shaming is a valuable currency: "She didn't go to church on Sunday. She must be a whore." Being a well-known person is the same kind of thing, it's just on a very big scale. If you yearn to be famous, think on it long and hard because it is most definitely not all that and a bag of chips. It is a hard life for the mind. It is about so much more than you can possibly know, and it's all meaningless at the same time. A natural-born mind fuck. Remember what I was telling you about cults? Well, small-town cult mentality is the same as Hollywood's—the only difference is that Hollywood is filling the minds of the small-town gossips, giving them their belief systems.

Back to Manson times: In April 1999, the first mass shooting at an American school had just happened, a violent tragedy at a school called Columbine. Two murderers in the first televised high school shooting went on a rampage. When it was going down and those poor kids were being held hostage by the killers, all anyone knew was that the bad guys were wearing black clothing. CNN and other news networks immediately started putting out pictures of Manson, saying the shooters were Satan worshippers and fans of Marilyn Manson. All while a hostage situation was in play.

In their manifesto, the killers actually wrote about how much they hated Manson. They thought he was weak. But in the media, the damage was done. CNN, Fox, all those places, trying to throw any garbage on the screen in absence of coverage because they couldn't get inside the school building, put Manson's face on-screen and he became deeply intertwined with that horribly tragic day.

After Columbine, Manson started to receive death threats and bomb threats, and by association so did I. Manson's part-time security became full-time. The public at large was something scary and to be kept away from. The blame wore on him. All this enmity, and hatred and rage.

I'm so protective of people I'm with, and it became almost a full-time job being with him. It didn't leave me time to think about myself, which was both a positive and negative. Looking at the past, I can see why I did what I did. It was the programming that is given to us girls from the first, the programming that says men's careers are more important than ours. I tended to go out with powerful men who treated my career like a hobby whereas theirs was the real deal. I think a lot of women do this, too. I hope you look at your own life and recognize whether you are minimizing it, because I did. I hope you don't. Know that your work is equal even if your pay is not. I wish I'd been able to see my strengths earlier, but . . . I wasn't aware enough yet.

Eventually, I grew exhausted. The circus that saved me was now burning me out. I was really in love with Manson, I just couldn't do the lifestyle anymore. I was too tired.

I didn't ask Manson for any money. I didn't take anything, other than the furniture I'd already owned.

After we broke up in 2001, Manson went on Howard Stern and trashed me badly. I was so shocked. This was a man I'd nicknamed "Doctor," because if I even looked like I had a headache, he'd say, "Do you need aspirin, Tylenol, are you okay, can I help you?" He was so considerate of me, so gentle, but he became someone who had this vendetta, and it was nasty, hurting me badly. He always acted like he was so different, but in the end behaved like a typical cisgendered male, that is, harassing the defenseless woman because his man ego is hurt. Waaaaah. Poor wittle baby.

How about you grow up, boys? That attitude leads to our death. We can die in many ways, it's not just the body.

After the breakup I resurfaced in Hollywood. Because I had to make money, I felt as if I had to go on a campaign of "See, America? I'm not that scary!" I had to do that just to survive, because what other kind of job was I going to get at this point? I was infamously famous. I was stuck. And I knew I was being blacklisted in the film community by my assaulter. So what were my options?

ON SET

I wake up Monday, at 4:15 a.m. Is it the morning or the night? I'm not sure, I only know being up this late/early makes me nauseated. The night before, I memorized all my lines, around ten pages of dense dialogue. It is tedious going; the words on top of words on top of words—all varying just slightly from the day before—make it difficult for me to remember. I have dreams that I forget all my lines and everyone just stares at me in silent expectation. I am continually stressed I won't remember my lines; each day that I don't forget them feels like a huge victory. Time is definitely money on film sets, and anytime I mess up I can feel the stress of the assistant director and producers, which makes me more nervous; everything hinges on us actors not fucking up.

I drive in the dark to the set and arrive. I'm immediately followed by a production assistant with a walkie who radios out that I've arrived. He follows me to the hair and makeup trailer. They follow me everywhere all the time—these guys with the headsets and walkies—as if I'm not capable of walking from point A to point B. I step into the trailer and my eyes adjust to the bright lights. The blow-dryers are already going, but even so I can still hear my costar screech-talking in her baby-talk voice. I put my earplugs in to block the cringe-inducing sound, but I can still hear her squawk. I hate baby talk unless there is a baby involved. My makeup artist remarks that I must've slept well because the skin under my eyes is puffy. She tells me every day how much better I look when I don't sleep. I stare at myself and my under-eye bags look not too hideous, but if she says they're bad, it must be true, I guess. I hate sitting and looking at myself in the mirror. I think it's unnatural to have to constantly stare at your own face with a critical eye. I stare trying to recognize the person staring back, because this is where I morph into another woman, not myself. The fucking baby talk is still screeching and I shudder. I get sent over to the hair chair and the blow-dryer begins its forty-five-minute job. I have a lot of hair. The hairstyle I'm being given has to match the picture of what was filmed the day before. I read over my lines on my miniscript; yep, they're just as banal as I remembered. I'm finally done with my transformation. I survey the finished work. I'll get to be myself again in twelve hours; until then I have to be someone else, a passenger in my own body.

Funny thing, actors, a lot of them; we spend a lot more time living in trailer homes than most people would imagine. The lifestyle is not as glamorous as you might think. The tedium can be oppressive.

Take filming, for example. Imagine the scene is you having coffee with a friend, just the two of you sitting there in a coffee shop. That two-person scene will take about four hours to shoot, if not more. You need the master shot of the entire scenario. Then you go in closer. That's a medium shot, and then we find our heroes. Now, there's a medium close-up of the hero and then a tighter close-up. After that side is complete, the camera turns around to the other side and repeats. It's a lot of work. You say the words over and over and over again from every possible angle and that's just one take. You have to do many more takes on top of that. I'm not complaining, I'm explaining.

You start getting very, very tired around 4:00 p.m. because you've been up since around 4:30 a.m. You cram in memorization for the next day's work. The day wraps up and you go home at 7:00 p.m. if it's a good day. Often it's around 9:00 p.m. The 9:00 p.m. means you have what's called a turnaround. You have to have twelve hours in between shooting times legally, but there's something called forced calls where they can make you come in after a ten-hour turnaround. That's what usually happens. Now you have about an hour or so to be awake and be yourself. You look at your lines again before bed, wash your face, pass out, get up, start the whole thing over again.

TELEVISED LIFE

I was in Romania when I got a phone call from Aaron Spelling, a legendary TV producer, famous for a slew of wildly successful if cheesy shows like *The Love Boat* and *Dynasty* and *90210*. He told me about a show called *Charmed* that had been on for three seasons; it was about good witches who helped protect the world from evil. There had been a major cast shake-up. Shannen Doherty, the show's main star, was leaving the show and they were writing in a new character. Would I consider playing a role on television? I didn't know what *Charmed* was, but I was curious. On the flight home from Romania, the pilot episode of *Charmed* was available to watch. I have never seen *Charmed* offered on a plane since then, just that one flight. It felt like kismet. I'd been looking for a sign, and there it was. I thought the pilot episode was pretty cool so I agreed to meet and talk. It felt nice to be wanted after so much rejection and knowing I was blacklisted by my omnipresent Monster.

The bigwig's office on Wilshire Boulevard in Los Angeles was like a football field of deep white shag rug. So deep I couldn't see my high heels. A tiny little man, probably about five feet tall, was sitting behind a big desk, attended by his butler in tails. This was the legendary Aaron Spelling. It was wild because I used to love watching *Dynasty* with my mother when I was little and here I was now with the head honcho. The butler brought Mr. Spelling this big crystal goblet of blue Gatorade with a bent straw on a silver tray. I'll always remember that. I imagined the conversation with Spelling: *I like a bent straw*. I just love that.

He was a sweet man. I didn't dislike him; in fact, I thought he was kind of cute. The whole time I was in the meeting, I was imagining him with blue troll doll hair. That's what I often did so I wouldn't be nervous.

I started to speak and noticed Mr. Spelling looking at me like I was part alien, and so did the couple of other executives who were in the meeting. That always happened everywhere I went in Hollywood. I think it was because the way I looked and the way I spoke were two very different things. Meaning I looked like a stylized bombshell, but definitely did not speak like one. People in Hollywood and the media world really did not know what to make of me. Instead of seeing me as an artist and mind, they put me in the "bad girl" category. They couldn't seem to hear a fucking word that came out of my mouth because they were too fucking stupid. Tedious. I'm like the most patient impatient person in the world.

I was fine with being known as a rebel, but worldwide condemnation of you for being a bad girl will seep into your head. Plus, the "you're bad" thing triggered a lot of my father's earlier words and made them echo in my head. The old fear of being homeless, being helpless, began to creep back. I was scared of television because I'd heard that hours were brutal, and I didn't know anything about that world. I barely watched TV. But it was a job. There was about to be an actors' strike in Hollywood. I knew there might not be work for the next two years.

I started to think seriously about doing the show.

My movie blacklisting was still hanging over me. It felt violating that so many industry people knew of my trauma, but sided with my perpetrator. Years had passed since I had last had regular work in Hollywood, and I was only twenty-eight. I remember being embarrassed on my birthday because *Entertainment Tonight* announced my age. I felt ashamed. That's how fucked up my brain was. The messaging that you get from society, media, and Hollywood skews age-related thinking. Back then I thought it was just because young women were more beautiful, but now I realize that it's because a young woman is typically more docile, often easily manipulated, and too easily tricked into doing what the man wants, both societally and individually. Fuck that. Funny how when we come into our own power as women, that's exactly when we are considered unattractive. It is they who are unattractive.

So I said yes to *Charmed*. I was told that it wouldn't run for more than two seasons, so I may as well sign a standard five-year contract. I should've realized that it was in other people's moneymaking interest to do this, but not necessarily mine. Everyone was hoping I'd keep the show on the air and get them to their ultimate moneymaking goal, the Holy Grail known as syndication. That's a hell of a lot of pressure for someone who had only been in indie films. I now had 150 crew members' jobs to think about. Believe me, it factored into a lot of my decisions. For the next five years I was my character, Paige Matthews, more than I got to be myself.

Paige Matthews was half sister to the Halliwells, Phoebe, played by Alyssa Milano, and Piper, played by Holly Marie Combs. Alyssa Milano I'd known about because of her show *Who's the Boss?*, which I was a fan of when I was small. I was somewhat surprised to find myself in a world where we were together, we were so totally different. Her, raised as a wealthy child actress, and me, the scrappy weirdo. There definitely could have been drama on the set, but I refused to take part in it. The press was salivating at the idea of juicy tabloid stories coming from our set, but I refused to oblige. My interest in the job was purely mercenary, so I opted out of infighting. I refused to play games that were beneath me, but it didn't stop the constant watching of our set. I cannot tell you how many times I was asked if we the Charmed Ones hung out after work. The answer was no. The one free hour I had when I went home for the night I tried to spend with myself.

I channeled my efforts into playing Paige, a lovely character, but very much a character. A girl who grew up into a woman on the show. Meanwhile, my own growth as a woman was stunted because of it.

I had so many life milestones on camera instead of in real life. My "first" wedding was a very meta experience. I walked down the aisle with a fake dad, with fake friends, fake sisters, a fake husband-to-be, a fake pastor: it was all fake fake fake. I went through the emotions as if it was really happening to me, because that's what acting is after all. It was particularly weird faking a life event that has been sold to all of us as some monumental moment—"the most important day of a girl's life." I don't necessarily agree that your wedding day should be the single greatest event in your life, but having it happen on-screen for the first time robbed something from me. All told I was fake married three times on film before my "real" marriage. By then, I was repeating an emotional scene I'd already played. Your entertainment comes at a cost to us performers. You should know this and acknowledge.

When I started the show, I thought a lot about audience psychology. I, in all my infamy and not-girl-next-doorness, was supposed to step in after Shannen Doherty's beloved character was unceremoniously killed off. I wasn't going to be playing Shannen's role, but it was still someone new for the fans to bond with. I had been told that many shows don't survive a major cast change. I knew that I had a slim chance at success here. People had to fall in love with my character as quickly as possible or *Charmed* would die. I thought about how big the crew was and how they would all be out of a job if I failed. So I made myself look super nonthreatening. I gained weight, about ten pounds, for the role. I wanted to look as soft and approachable as possible. Boring.

There was an edict from the studio that none of us was allowed to cut or color her hair without permission from the studio president. I thought that was some serious bullshit. As I was about to return for my second season on the show, I dyed my hair red without asking permission. Oooh. The studio got wind of it and flipped out, of course. They were furious and demanded to know how I expected them to explain this. I told them: "It's a show about magic. You simply say I was mixing a potion and it exploded in my face! My hair turned red! I liked it, so I kept it." That became the very first dialogue of the season, almost verbatim, between me and the character Leo. All it takes is a little creative thought, but the studios rarely have that working for them.

There was another big to-do on the set when Holly Marie Combs cut her bangs. The studio feared people wouldn't recognize her. That's how dumb they think you audiences are. If a character changes their set look in any way, audiences won't know who they are looking at. I wish you could all be flies on the wall when studio executives and producers talk about you all. Believe me, you are looked at as sheep with no minds of your own.

I'd become almost a sheep myself, and I was losing my

mind. Day after day, month after month, year after year, my brain space being eaten up by dense dialogue that's just slightly different from the previous episode, will do that to you. Eight to ten pages a day. I was having the same conversation that I had had in the episode a week before, but with the guest stars slightly altered. You could've interchanged anything I said in almost any episode with any other episode. That was the part that just made me insane. It was prison for my mind. I was starting to lose my grip on my sanity. And I was so very lonely.

Sometimes to amuse myself I left the set at the end of the day with the fake blood still on my face or a giant wound, and I'd stop at the grocery store on my way home. I would do it to test the public, to see how they reacted to a wounded woman. Nobody asked to help me. Not once. They would just avoid eye contact and look down at the floor. It gave me great insight into the invisibility of an abused woman. Note to society: you should always ask if a woman needs help if they look abused. Do the right thing. Be bold and be brave.

I've always liked to conduct social experiments. I think it was my way of keeping sane and entertaining my mind, which was just atrophying during this period. I could feel my mind dying, and I didn't know what to do. I was too tired, too lonely, and too sad to do anything about it.

I made it a point to not involve myself in petty set politics. I think sometimes when people have led coddled lives, they like to create drama—it makes them feel like something's happening, something's going on. But I had had real drama, real problems. I refused to create nonexistent problems to give my life a sense of faux importance.

Only one female director was hired in the entire five years I was there, and the crew sank her. This was a show about three young women, and they had not in all the time I was there had a female director. But the mostly male crew, I think without even realizing what they were doing—they just cut the female director's legs out from under her. The crew would snicker in disrespect when she would direct them. I can't imagine it was a pleasant working environment. I feel horribly about not fighting for her more, but I didn't fully understand the dynamics of what was happening. My character was too busy talking to leprechauns to have the time.

Anytime on that show that I said I really liked a director, if it was someone from outside of the Spelling all-average, all-white male mob, which was very rare, they never returned. If I said I hated somebody, they would return for sure. There was one director we had, whom I was so excited to work with until I met him. The day he showed up to direct we were in my fourth or maybe fifth year on the show, and the show could direct itself. It was a well-oiled machine.

So in the first rehearsal of the day, after walking from point A to point B, I exited right. It's not a hugely important decision, and the new director hadn't given any direction as to whether to exit right or left. But he exploded. He yelled: "You idiot. You do not cross in front of my camera! YOU DO NOT CROSS IN FRONT OF MY CAMERA!" He kept yelling. The man had a meltdown. His voice reverberated off the walls. I was in shock. He kept yelling and now started cursing me out. Bitch. Idiot. Bitch.

This was exactly one minute or so after we said good morning to each other. It was 7:30 a.m. No one shut him down. Not one person on the almost all-male crew, not one producer, not the assistant director. Men I had worked on the set with for years. Why? Because the director, even if there for only thirty minutes, was a guy and I was a young woman.

I was thinking, *Excuse me, I don't even fucking know you. You're in my house. Yet you feel so comfortable and assured in your place in this firmament of male directors that you're going to do this to me?* I told the producers: "I will never work with that man again." They assured me he wouldn't be allowed to verbally abuse me again.

Guess who was back two months later. This is just a minor note in the screamer directors I've endured. Hollywood is known for screamers and belittlers, and their abuse gets a pass. I've never seen or heard a male actor be abused on a set. It's time for people in any workplace to do what they can when another is being bullied and terrorized; it's time to stand up for those around you. If you do not, you will be called out, you will be shamed, you will be dragged.

It was just another example of a man getting a free pass. Can you imagine in the first hour of your new boss's day he screams at you in front of everyone and not one person steps up? Maybe you can. I hope not, for your sake. There is no excuse for abuse.

Not ever. Being screamed at aggressively is something the body takes in and it's jarring and traumatic. But you know, go put your short skirt back on and say the lines, little lady, 'cause you're an actress so you better take it.

Yes, there are other industries where there is screaming, but you have to understand, there is zero federal oversight in Hollywood.

It's hard to reconcile the sweet little old man I met—Aaron Spelling—with all the decades of sexist media he'd produced and put out into the world. Credit for that sexist media also goes to the people he surrounded himself with: the executives, all male; the writers, mostly male; the head writer, male. On a female-driven show, of course. What made all of the sexism on the set even more galling was the fact that the show was carried by women, made for a female audience. Because they know how to get into young women's minds, for sure. There is a completely understandable outcry over whitewashing, casting white actors to play roles that were in the source material meant to be for a person of color. Why is there no outcry over men telling women's stories? There should be; it's not like they do it well. They do it so it's just good enough. I'm bored of good enough and I know the public is too.

During year two of *Charmed* I decided to go to hypnotherapy. I found the repetitive days so opposite to my natural rhythms that I became sick over and over. And it was at times a very stressful environment. I started to have panic attacks because of everything I was pushing down. We filmed in a place called Woodland Hills. Simply seeing the word *wood* was enough to make my heart race and my palms go sweaty. I was sick about four or five times a season. We would shoot twenty-two or twenty-three episodes. On hour-long TV, you are essentially shooting half of a feature film in eight days. The pace was grueling. Two years in a row I had 102-degree fevers and got dropped into giant trash cans, in a stunt, always on the days that I was the most ill. That is the only time I allowed myself to feel sorry for myself. Part of my being a hard-ass on myself for my whole life really kind of hurt me in the long run. The exhaustion was real and raw. Twelve to sixteen hours a day for about five years straight in a wash, rinse, repeat environment unlike any I'd known. Kind of like being in a really fancy factory, like working on an assembly line of sorts.

One day, three of the *Charmed* producer men strolled up to my trailer to inform me that my one friend on set, Sam, had been fired after being caught smoking weed on a school bus set. Sam was cool, not the usual Spelling crew member, with his dreads and his real smile. I was genuinely saddened because Sam was the only one I could sense was like me. But I said it was for the best, his soul needed to be free. I couldn't help smiling as I asked the producers, "So if I smoke weed on the set, will you fire me? Could I be free?" Their response wiped the smile off my face. "We will garnish your wages for all time, no matter where you go. We will take your money, and we will ruin you and you will never work again." All said with extremely straight faces because they were not fucking around. Fuck. This Hollywood detour of mine took me down the wrong alley. I sure did meet a dangerous crowd. Only these were rich white men on studio lots, not in a back alley.

It's a particularly terrifying thing to have men tell you that you're going to be punished so badly you will never be hired again. But that's the pervasive attitude in Hollywood: "Don't step out of line, little girl. We can just point to the next one right after you." There is a constant stream of young and damaged women flowing into Hollywood.

If you are reading this book because of *Charmed*, thank you for being a fan of the show and my character. I respect you and I honor you. I know it's brought a lot of joy to a lot of people. I was glad to be of service. I am glad something really good was coming from what, for me, was a very difficult time. When I say anything that's negative about the show, you have to understand I'm speaking about my personal experience. Not Paige Matthews, my character, but me, Rose. But you can always find good in most situations and so that is what I choose to focus on. There is much that I'm proud of from this period. For a long time *Charmed* was—and might still be—the longest-running female-driven hour-long show in history. I wish we got more credit for that, because it's important.

In 2006, as *Charmed* came to an end, I returned to a changed media world. When I left society five years before to do the show, celebrity culture was different. The gossip websites now ruled. The paparazzi were now insanely aggressive. The tabloid magazines were drunk with profiting off Lindsay Lohan's antics and Britney Spears's troubles. And if you were well known, you were hunted, not just by the paparazzi; the public at large became something to fear. Every stranger with a phone became a potential informant. It was the first time people started posting online where you were, what you were saying, what you were doing. Everybody was reporting on anyone well known and you didn't know where the enemy was because it was no longer just the paparazzi or the gossip columnist. It was everybody. I was forced to live a life where I didn't say anything, I didn't do anything, lest I be branded a bad girl. Most were so freaked out by people reporting false items, they just stayed hidden. At least that's what I did.

I was overwhelmed by the media's new world order. There was no Instagram to speak for yourself, so if lies and smears were written, they were taken as fact. And, wow, was I smeared.

Even in a place that's supposed to feel safe—your work environment—I felt hunted and objectified. Just walking back and forth to the set on the Paramount studios lot was stressful; there'd be a long train cart, packed with tourists, all pointing at once and taking my photo. It was like: "There's the actor in her natural habitat. Please don't feed the actor." To be the one on the other side of the hunter's camera felt vulnerable as hell.

I decided to escape America for a bit, thinking the hunting would be less prevalent. I got chased out of the Vatican once by about thirty German tourists. I shouldn't have been at the

Vatican that day, I was so sick, but I had desperately wanted to see it. I clung on to the old stone buildings trying to get away, and then finally, the crowd caught up to me. I was cornered. I flattened myself against the wall. I remember thinking, *We're at the Vatican. Isn't seeing Michelangelo's work more important than someone on television?* I was sweating and scared. I was being jostled as someone pushed through the crowd, I thought to help me. Instead a man grabbed my hair and yanked a small clump out. I screamed and pushed, but I wasn't being let out of the circle. When they surrounded me, it was like a wild pack of dogs. I couldn't see their faces because they had their cameras pointed at me, the flashes blinding me. I could only see hands as I got pushed sideways and back, white circles where people's leering faces were. I felt as if I were being eaten alive.

People need to respect boundaries. Just because you see someone you *feel* you know, does NOT mean you get to harass them or touch them. The person you think you know owes you the same as any human does, which is nothing more than politeness. So be polite, be respectful, do not surround, do not grab, do not clutch, do not push. Be civilized and you will usually get that in return.

What I experienced in that crowd of "fans" was traumatic and assaultive. I had a small bald patch where my hair had been pulled out. These kinds of situations happened frequently.

Fame is something so many think they want, but the reality is something hard for them to see. The next thing many say is "you made money" as if money equals happiness. As if money equals safety. As if money equals peace. I would also say do not make your value system mine. When I finished *Charmed*, I was told by my then publicist: "If you need to step out the door and there's a camera anywhere, you better have someone come dress you and do your makeup." God forbid you show your actual style. God forbid you go to a store like every other person and put on something that represents who you are. God forbid you wear something more than once like a normal human does.

Hollywood publicists, whom you're told you must have and who cost around $6,000 per month, tell you what you should act like, look like, talk like—how you should respond to the press. They prey on fear and manipulate actors by threatening: "Oh, the studio will be really upset with you if you don't do this." They are a key part of the machine that keeps you in line. The kicker is you're paying them to fuck with your head.

In classic Hollywood times, publicists were the people who covered up what the stars did. Now, the stars don't even really do that much, not like they did back then, because they've been so trained to not do anything. God forbid the public find out you're human.

So in any case, after *Charmed*, my very expensive publicist told me I needed a very expensive stylist. I should've asked, *Why am I paying somebody to put their idea of what I am on me? I didn't have a stylist during my pre-*Charmed *era, and I seemed to manage fine; what has changed?* But my brain had been twisted for so long I no longer had an internal compass.

When a stylist is paid to come put a look on you, their main goal is to make you attractive to the male species. It's a very stereotypical idea of what is sexy, what is hot, what is fuckable. But it is not cool and, boy, did stylists do a number on me. I was made to look like a plastic weirdo. And I paid a lot of money to look like a plastic weirdo.

The bizarre thing about the industry is: you were discovered because you were unique. What was special about you when you were discovered, they do their damnedest to remove, not unlike what traditional society does to children. When I was discovered, I had a raw quality and my own style; at the end of my career it was Barbie hair and beauty pageant looks. Then, every photo of me was photoshopped. Meanwhile everyone online was telling me how ugly I was. Then I'd go to work and play other people. You need all your faculties to be in a semi-denial stage in order to survive it. Things that are abnormal become the norm, up is down.

And that is how I got lost from myself. I bought my first piece of "important" art at the height of my "fame." The piece is of a woman who looks like a ghost, which represented what I felt like at the time. Just lost behind this haze of objectification and misunderstanding.

I had gotten lost from myself, terribly, terribly lost.

DESTRUCTION

One night in Cannes, France, wearing the most beautiful Dolce & Gabbana red lace dress, I went to a party. A huge, glamorous party where I knew no one. I ended up sitting on a couch by myself because I tend to get shy in large crowds. I noticed a man sitting on the opposite couch, also by himself. He was wearing a black suit and a black cowboy hat. We were both there for some time, the two loners at a party.

There are some moments that you regret for the rest of your life. This is one of them.

I think I made a joke about how we were two fancy-dressed losers sitting alone at a party, and we both laughed. We started talking, and I realized he was a famous director. Goddamn, he was handsome. He had many hit movies under his belt and at the time was riding a career high.

He said, "Come sit next to me on the couch." I said honestly,

"No, I don't want to be seen as an actress sucking up to a director." Much like how a cult works, in Hollywood you're not meant to talk to someone whose status is higher than yours. And even though I was a star of a huge hit TV series, I had to watch out, lest I be branded a whore actress.

He asked me my age and I fibbed about it; I think I told him I was twenty-eight. I was thirty-one at this point. I was deep in the grips of Hollywood conditioning. The thing is, I was always playing roles that were younger, at least five years younger, which amplified my twisted perception of aging. You have done something wrong! You have lived! You start feeling crazier with each birthday that passes.

For me it was Fear with a capital *F*. Fear of the Future. Deep down, I was scared of where I would be in life at whatever age I wasn't yet. I was so fucked up by the Cult of Thought in Hollywood that I channeled my fear into trying to control the aging process. I wanted to freeze myself in time. Like so many in Hollywood. And it wasn't just my head that was telling me to; it was makeup artists, hairdressers, stylists, agents, managers, you name it . . . They want you to stay as young looking as possible for as long as possible so they can make more money off you. The human woman me has her own aging crap to deal with, but the actress me had a literal multibillion-dollar corporation built off her face and body. People are depending on your face to put food into their mouths and pay off their car loans. Women in general have the media world telling them not to age, but we in the public eye are being handed direct hotlines to people who can supposedly freeze us and are very encouraged to do so. And we have the money to do it. And we so often hate ourselves; our photos are as manipulated as our minds. We get the message we are not enough times a million. We are marketed to you the public as something to be envied, our looks that make you love and hate us. In the long tradition of Hollywood, one of its messages is "Don't you wish you could look like us?" At least until we go overboard with cosmetic fixes, then you start clucking and making fun of us. Being an actress, you are in front of a mirror for far more time than is healthy; having to consider everything about your face in relation to it being okay to be seen by the entire world is definitely not healthy; to have your teeth be five feet tall is not healthy. Ugh, it'll all fuck you up. I know, it did me. There are reasons actresses go overboard; for me it was like cutting. I hated what I had become and I wanted to destroy it.

For four years I barely had contact with anyone other than those on the set of *Charmed*. I was not in my right mind. I didn't know myself anymore. I'd lost my epicenter. I was rudderless. I was a prime target for abuse, but I didn't know that then.

That night at the party in the South of France, I didn't know anything about this director, other than that he directed those movies. After a while I went over and sat next to him. We instantly connected. There was something electric about that night.

This man would become the most significant relationship in my life besides my father. And later he would outdo even my father in cruelty. But I didn't know that yet.

The day before I'd met him, I'd been on the plane bound for France. I was bored, and the lady next to me handed me a women's magazine. I had a flashback to the days when I had consulted those magazines to see what adult women did. In the magazine there was a quiz: "What kind of man do you want?" I took the test and at the end it pronounced that I wanted somebody adventurous, funny, wildly intelligent, et cetera. Cool.

When I met this man, I thought he was all those things. What I didn't realize is that these quizzes would serve better if they asked what you do NOT want in a partner. Because there are many, many things I didn't even know I did not want. Like the quiz, I had always thought more about what I wanted. Knowing what you don't want is more important than what you do want, because the reality is, we rarely dream as big for ourselves as the universe can; by knowing what your nos are, you get to the yes, and the yes is bigger than you maybe even imagined. It would have served me well to have steered clear of the controlling, manipulative, and violent.

Our relationship began very platonically. He told me he was married but unhappily. That he came from a very traditional family where no one had ever gotten divorced. He stayed up all night long working and would sleep during the day, when his wife lived her life. He said they had separate houses on a giant compound in Texas and would separate after the kids were grown. As it turns out, it's the whole rigmarole that married men always tell you, like a broken record. I just didn't know that then. I believed him. And the lost me fell for it.

He magnetized me, instantly. I thought he was so handsome. Another thing that attracted me to him initially was that he had this real childlike way of looking at the world. I never had that and it fascinated me.

RR told me I was the most beautiful woman he'd ever seen. He couldn't believe I was so smart and funny. He didn't know women could be funny. He couldn't believe I was an actress. He told me how special I was. It was so gratifying that someone saw me, or at least I thought he did, for assets beyond my looks, for not being the cliché of an actress: conceited, frivolous, dumb, all the things that a lot of actresses, especially aspiring ones, are accused of being and, sometimes, are. You say "actress," and people roll their eyes in Los Angeles. I did feel ashamed that I was an actress—I wanted him to see me as being on his level. I was so desperate to prove that I wasn't like all the people in Hollywood, because he told me how evil he thought they all were. But evil is everywhere, and just because he didn't live in Hollywood didn't mean he didn't benefit from Hollywood and its permission to men to behave horribly.

This man in the cowboy hat sold his lifestyle as superior to mine: he said he'd lived a pure life where he'd never done drugs and never partied or lived wildly, which is why he found himself in such a successful situation at his age. I believed him. In my head, his career was much more important than mine; certainly the world valued him a lot more. My career was just something I did, something I was, frankly, embarrassed about. I've recently realized that I was the powerful one all along. But most of these powerful men, well, it's unlikely they're going to help you see that. Theirs is the career, yours is the hobby.

For three weeks after Cannes, RR and I had no communication. Then one day I texted him, and he texted me at the exact same moment. This was something we did continually, from that moment on. It felt like a sign that we were incredibly, deeply connected. He would send me a T-shirt of his with his scent on it, packed in a ziplock bag, and I'd send one of mine back to him. We fell fast and we fell hard. It was exhilarating, and so much more interesting than just pretending; this was supposedly real life now.

Please take heed: when someone tries to insert himself into your life very quickly and rushes to tell you he loves you, that should be a big warning sign. And as much as you want to hear it, as much as you're hungry to be *seen* with a capital *S*, as much as you've been lonely and just waiting for this, understand that very often these are the men who will turn on you.

By the time RR revealed his true identity, I was so deeply in love that I couldn't adjust to what was going on. He was so amazing to me at the beginning, such a gentleman, that I figured it must have been my fault, that I must have done something wrong, for him to turn cruel, to stop respecting me. No. It was his plan all along, conscious or subconscious.

I've found that when someone tells you they love you too fast and overwhelms you and wants to move in right away, it's a trap. Know a man like that most likely wants to own you and control you in order to make himself feel powerful and significant. Know that things will change. Try to look out for the warning signs and the red flags because there are many, and women tend to really discount them. I didn't see what was plainly clear—see, it's not just society that gaslights you, it's ourselves. We need to protect girls from these kinds of situations by telling them from birth that they are worth just as much as boys. I wish I'd known that. That our careers and endeavors matter just as much, that we have just as much potential and power residing in our bones—if not ten times as much. If we raised our girls correctly and didn't fill their heads full of fairy-tale weddings and Prince Charmings, full of beauty magazine bullshit advice about what a man wants and how to please a man and how to keep a man, a lot of girls wouldn't fall for these dangerous men, wouldn't fall into these dangerous situations. We could save a lot of lives. As for me, I had never been wildly in love before, not like this, not out of control. I still had one final season to go on *Charmed* and was annoyed that work got in the way of us being together. We were totally engrossed with each other. I looked at him as my lover, my friend, my life. We were obsessed with each other.

We got together in my Spanish house, where we hid out from the paparazzi, the fans, the world. I couldn't go anywhere without being followed, recorded, spied on, at this point. I had lived in my own bubble even before he showed up, so it was easy to keep what happened between us in my house secret. He was very concerned about his public image, and he loved being known as the guy who stayed together with his wife. One of my biggest regrets in life is not comprehending the damage we were causing, not comprehending collateral damage, not comprehending my damage.

His wife and kids were in the faraway state of Texas and easy for me to ignore, just like he did. I didn't mention their relationship. He didn't, either. In my head it didn't exist. I didn't want to know. I took my cues from him. He would tell me things like he had moved out and then gone back to her. He had gotten married at nineteen, too young to know better. He made it sound like the only thing keeping him with her was the fact that no one in his Catholic family had ever had a divorce, and I thought to myself, *You just need to be a braver man. Why would you live your life in a way that wasn't true to yourself?* This was very selfish of me. But also, I think he was living true to himself. By cheating on her. But the me-me wasn't really there at this point. I was so scared all the time. Scared he'd leave, scared the man I had waited for would go. I had no idea what I needed at this point. Yuck. I was needy, a broken person, and I felt whole with his love shining on me. I am embarrassed by all this now. There's an old song that goes "Looking for love in all the wrong places." Yeah, that about sums it up.

I found out too late that to have a secret relationship really suits a manipulator. He encouraged me to cut off all contact with my friends, not that I had many locally in Los Angeles, and with my family. Nobody could know about us. I simply disappeared from the outside world, even more than I already had because of *Charmed* absorbing my life.

This relationship had all the hallmarks of my mother's relationships. And I thought I was soooooo different. Nope, even worse than not different, I was now a cliché. I wish I could go back in time and get myself the hell out of there and not be involved in hurting others. Nope, I was in for five and a half years of gaslighting.

Some months into our relationship he told me he was writing the lead female part for me in his next movie, which was going to be a double feature with Quentin Tarantino. They had both grown up in the 1970s seeing these pulp movies called grindhouse movies, which played in theaters as a double feature. They were played together with these fake trailers in the middle, as an art piece. They were best friends and pigs in shit. RR's film was going to be called *Planet Terror*, which in retrospect is the perfect title for a lot of my life with him, and Tarantino's was to be *Death Proof*. They were salacious, but as female-exploitation flicks go, they're pretty great art; they're punk and fucked up. But yeah, objectification was on high. And so was intense abuse of women, both in reality and symbolically.

RR started coming to visit the set of *Charmed* at Paramount studios, and we would meet up in my trailer. People saw, but because he was a writer-director who had started writing a part for me, it was okay to be semiseen in public.

I profoundly regret and publicly apologize for my part in this. I carry a deep, deep regret for the pain and heartache I caused. Also, not insignificantly, I could have saved four years of my life and my own heartache and pain. But at the time I was warped in the head.

As I had with so many people, I told RR about my assault at Sundance early on in our relationship, the first or second night we saw each other. I still was consistently having nightmares where I'd sweat through the sheets and wake up yelling, and I was still shaking because I saw my Monster at an amfAR charity event. I was seated next to the Monster at the event. I met RR later that evening. I was in shock. RR of course said the traditional macho thing: "I'm going to beat him up." Of course he never did, never would: you see, RR was a faux tough guy. As most bullies are. But I didn't know that quite yet. I believed him and really looked forward to the day when my perpetrator would get punched.

I was really excited about working on the film together, *Planet Terror*, for a couple of reasons: one, because it sounded like a dope film, and two, I wanted to get back into film through the back door, in a way that the Pig Monster couldn't sabotage. I thought if I came out in a Rodriguez or Tarantino film, I'd be protected. I couldn't be hurt by the blacklisting anymore. I could finally work at a level that matched my interests and tastes. I'd been starved mentally for years, robbed of varied roles and experiences by a bitter relentless Swine. Speaking of swine, turns out Tarantino knew about my "settlement" with the Pig Monster before I even got cast in *Death Proof*. A movie he made me audition for three times. The first time I met Tarantino, and for years after, every time he'd see me, he said, "Rose!

I have your movie *Jawbreaker* on laser disc! I can't tell you how many times I used the shot where you're painting your toes! Your feet are in tight focus and your face is blurry. I've used it so many times!"

Let me break down what Tarantino is *really* saying:

"I have your movie *Jawbreaker* on laser disc!": He went out of his way to buy a collectible, at least ten more dollars for the laser disc.

"I can't tell you how many times I used the shot where you're painting your toes!": Tarantino has a known foot fetish. To him seeing a naked foot is the equivalent of a breast person getting turned on by nipples.

That means Tarantino paid extra money to jerk off to my young feet and told me about it loudly, over and over, for years, in front of numerous people, as if I should be so thrilled that he donated his solid-motherfucking-gold semen that is clearly better than all the other semen in the world and he gave it up for little ol' me? It's time men realized their semen isn't all that.

RR said he was going to be my savior in the film industry and I believed him. He was my knight in shining cowboy hat. He drilled it into my head over and over, how lucky I was to be with him. I'm mortified that I had such deep programming—the man-as-savior thing—that I bought it. But who was buying it? Me, a lost and programmed Hollywood cult member.

RR was wildly jealous, which at first, like most, I took as being flattering. I took off a turtleneck sweater one night to show him my new bra and he asked me, "What kind of message are you sending with that bra?" I froze: "What? What are you talking about? I'm wearing a bra under a turtleneck." "You're sending a message to men in that bra." "What? No, I'm not." A huge fight and tears ensued as I tried to prove I wasn't trying to lure men to bed with my bra.

If he saw an old photo of me with any ex-boyfriend, or heard any reference to one of them, it would set him off into a rage where he'd do something utterly brilliant like throw his laptop out the window. Obviously because of the press he knew whom I'd gone out with in the past, and it drove him crazy.

When he got angry, his eyes turned black, even under his eyes. Like my father's did. He's a big guy with big broad shoulders and when he was mad, it was beyond intimidating; it was terrifying. When he'd rage, it felt like my hair was blowing back from my head with the force of his anger. My hair blew back a lot. I took to hiding more of myself to avoid explosions.

It became a steady cycle of being accused of imaginary sins and me putting out fires constantly so he wouldn't blow up at me. That's how it is. You start hiding some of your past, doing all sorts of things to prove that you're not what you're being accused of.

You're in reaction mode at this point. The gaslighter does this to keep you destabilized because if you're always reacting to a flare-up or an explosion, or getting berated and you just want to avoid it, you start monitoring your words, monitoring everything, hoping pictures of you alongside a guy from your past don't surface because there will be hell to pay.

He could be very, very vicious and very damaging. Something he said to me over and over was: "I got you at your ripest." I got you at your ripest. Meaning he got me at the perfect age for beauty. That is so disgusting and so damaging to a girl's brain. It filled me with anxiety and panic about aging, knowing that I was not going to be "ripe" for long, that I must do everything I could to stay this perfect being. Because if I wasn't perfect, then what was I?

It quickly became normal, being stressed all the time, because I was still so in love. And intermittently RR would make up with me, of course, and be amazing for a day. I would see this purity of soul, this luminous and magical thing inside, so I wanted to catch it like a firefly. But then that firefly turned into Godzilla and breathed fire on me.

When RR later asked me to marry him, which was so monumentally absurd, I didn't know what to do. All that was going through my head was, *Don't ask me in Texas, I fucking hate Texas. I don't want to have this memory here*, but he did ask. And I said yes mostly because I didn't know how to say no.

My main reaction was thinking I didn't want to hurt his feelings. I didn't know how to say no, because I never saw that reflected anywhere in popular culture. You virtually never see a woman saying no to a marriage proposal. All you see are girls freaking out, so overjoyed that someone finally wants to marry her. A man chose her. Oh yay.

I think politeness is a particular curse for girls and women. That's what *Dawn*, the short film I directed, was about. My direction to the male lead in *Dawn* was simply: "Basically, she's a mouse and you're the king cobra and you're hypnotizing her so she doesn't ever even see what's coming and you are going to do it by scrambling her brain." It's been done to me by men and it was done to me by Hollywood. It's what we do to girls. We send them out into this world with politeness as the strap that keeps our hands tied behind our backs. Then we meet wolves. It kills us. I read about a girl—her male neighbor kept offering to help her carry her packages, and when she repeatedly refused, he accused her of being the kind of girl who refused help. She finally relented and, of course, she wound up being attacked. That's what politeness does.

In any case, I said yes to RR, even though I didn't see us married. I did still love him at that point. The media went crazy over the engagement, of course. Perez Hilton, that oily gossip columnist, started writing "WHORE" across my face every day, but not RR's.

We were living in his castle—seriously, he'd bought a place that the previous owner had turned into a castle, a former water tower, embellished with turrets and everything. RR had commissioned a gigantic painting of me on the ceiling above the stairs. I was nude. It's pretty weird to have a six-foot naked version of yourself above your head. I was caught up there, at my ripest.

He played increasingly cruel power games. I had told RR I'd always seen myself having a little girl and naming her Cherry Darling. I never exactly saw myself having a kid with him or with anybody specific, but this daughter was real enough in my mind that I had written letters to her, stuff that she could read from me when she was older. And RR said, "I'm going to name the character I'm writing for you Cherry Darling. That way you can't have a child with another man and call her that. This is our baby. It'll last for all time. This is the child we're making." That really should have been a breaking point. His taking that name for the character was a serious form of theft. Because he was right: after that, I would never be able to name my child Cherry Darling.

As confused as I was, the role of Cherry Darling was shaping up to be amazing. One day early on he called me and said, "Machine gun leg. That's it. She's got a machine gun leg," and I was like, "Cool, right on!" I did love his imagination. The role was written for me, with a lot of the things I actually said and did and a lot of passages that are direct quotes from me. I think it's probably his most realized female character, one not just there as a sexy caricature. Even so, on the page Cherry was more sex fantasy than human being: I mean, she was a go-go dancer. But I fought for Cherry. Cherry, in fact, saves the world.

Tarantino's half of *Grindhouse* was *Death Proof*, about kick-ass cars and women dying horribly.

What's done to the women in *Death Proof* is disturbing. Hollywood is very comfortable with sanctioning the abuse of women and calling it art. I don't think most people understand that when audiences eat this shit up, it's programming them by desensitizing them. They learn to only really see women in an objectified sense. The poison carries through the audience. The audience then goes home and mirrors that behavior. And I'm telling you, what goes on behind the scenes bleeds through to you.

I feel that when you see how women are treated on-screen, you see what the director really thinks about them. You see how little he values them. Tarantino's always lauded for having strong women characters, but I would say look at what they go through. He beats the shit out of them for enjoyment. Yes, you had Zoe Bell, and Tracie Thoms, and Rosario Dawson kicking some ass, but look what's done to the other women. Sidney Poitier's daughter, named Sydney Poitier, played a character named Jungle Julia. Tarantino was always on her for not being street enough. It was so embarrassing. She dies by having a car split her vagina in half, her torn leg flying out the window. Another woman has her face skidded off by a tire. While shooting the death scene of my character, Pam, I got tortured in the car and had to shatter my face against the plexiglass. What the female characters in Tarantino's films get in terms of strength, they pay for in brutality. And once again, it's okay, it's just women.

I wanted *Death Proof*'s Pam to look very different from Cherry in *Planet Terror*. I wanted Pam to look like this angelic creature, because I knew she was going to get the shit kicked out of her and die in a violent and horrible way. I thought psychologically, *If I can make her look angelic, I can make the audience feel something for her before she dies.*

The filming began and the nightmare began. I was shooting *Charmed* during the day in LA for half the week, then I'd fly to Texas to film *Planet Terror* at night for the other half. It was toward the end of the entire *Charmed* series, almost five straight years of working those twelve- to sixteen-hour days I mentioned earlier. I was ground down and exhausted. But RR was constantly talking about how he hated actors who complained on sets. So I never complained. I should have. It was a brutal schedule. I would be on the airplane going back Monday morning at six after shooting all Sunday night, every time barely making my flight. When I arrived in Los Angeles, I had to rush straight to the Paramount studios and be a whole other person, this wildly different character, Paige. My brain was starting to scramble on top of its already fragile state.

Cherry was scantily clad through the movie, so I wore a lot of spray tan; it felt like at least something was covering me up. We used an aerosol leg spray. I learned about it from some of my friends who perform drag. It stays on no matter what. I'm probably going to get some crazy-ass disease, because in the small, enclosed space of my trailer, the makeup artist and I sprayed my entire naked body every single day. We went through a can a day, and these are tall cans. We shot in severely humid Texas, so it would never dry. I left orange smear marks everywhere I went. I would get on the plane to LA with my fake tan and a packet of baby wipes, and scrub my skin. I had to turn pale again to look like Paige. It looked like I was molting.

We shot *Planet Terror* at night, all night long. RR worked with more or less the same crew over and over, so they knew him very well. They stared at me since I was the outsider, and everyone knew we were together. And his wife—from whom he was now officially separated—was the producer. That just seemed cruel and unusual for everyone involved. I think that was a form of psychological torture for both of us.

RR's campaign of terror started the first night of filming. The character that Freddy Rodriguez (no relation) played was named El Wray ("the king" in Spanish), which is now also the name of RR's television network, El Rey. The character El Wray was a stand-in for RR himself. There was a scene where El Wray and Cherry (Freddy and I) kiss. For a week before we shot it, RR mysteriously started growing out his facial hair. He didn't tell me or anyone else what it was about. When it came time to shoot the scene, RR banished everyone from the set except for himself and me. You don't usually do that when it's just a kiss. He got a mirror out and shaved his beard into an exact replica of Freddy Rodriguez's, focused the camera on a super-tight close-up of just our lips, and made out with me on film. All this effort, so Freddy wouldn't kiss me. Then RR shaved off his beard really quick. He was clean-shaven when he opened the set again. Of course, nobody said anything like, "This is fucking weird. What the fuck is going on?"

The weirdness didn't stop there. RR had cast his nieces as "the babysitter twins"—they don't even get to have names. The camera pans up and down their bodies, ogling their breasts, sexualizing them completely. I think they were fifteen or sixteen at the time. In the scene the two young girls are lying on the couch and jiggling each other's breasts with their respective feet. This was their *uncle* directing them to do this. These guys don't care how badly they damage a girl's self-image. They feel like it's totally their right to do this. I was really grossed out, but I hadn't had enough distance from the industry, and certainly not RR, to piece it all together for myself. I just knew I was very uncomfortable with it and that nobody said it was wrong, but it felt wrong.

But the biggest mind fuck was yet to come. Knowing what had happened with the Monster, RR wrote a scene where Quentin tries to rape my character. I didn't even know how to articulate the wrongness, so I didn't. Maybe RR thought it would be cathartic for me? I did enjoy stabbing Quentin Tarantino in the eye with my broken wooden leg. But rape as a plot device needs to be shut down. For starters, it's lazy writing: there are plenty of other ways of developing a female character into a badass or motivating a female character to take action. Worse: it's exploitative. Even worse: it's retraumatizing to the women who have been molested or raped, and it's damaging to the actress, who has to pretend she's being assaulted while the cinematographer is on her ass trying to make viewers think, *This chick's hot*. It's nearly impossible to not feel like the violation is real.

Not long into the shoot, all these wild, inaccurate stories started coming out about RR and me, that we had just met on the set and we were fucking in the trailer. They ran on these gross sites like Perez Hilton and E! and papers like the *New York Post*. RR kept on screaming at me and accusing my friends of leaking these stories to the press. But I didn't have any friends. He had isolated me from everyone, including my family.

Of course in the media, it was I who wore the scarlet letter, not RR.

The shoot was physically grueling. I'm somebody who pushes myself, but RR pushed me harder, punishingly. I needed to run faster than everybody else, even though I was wearing a four-inch high-heeled boot on one side and an eight-pound cast as a stand-in for the machine gun on my other leg. My body has never been the same.

For one scene I had to jump over a one-and-a-half-story wall. I got lifted by thick wire cables, but I had to run as fast as I could (in my heels and cast) to reach a certain point to be able to fly. Right as I reached the top of this big gray cement wall, I had to be vertical, with my arms spread, and then I had to go horizontal, onto my stomach, perfectly parallel with the asphalt below me, and get my arms around these cables; otherwise they'd get ripped off. So I did it. I ran and hit the mark perfectly. My years of dancing really helped with a lot of the physical things I did in my movies.

I consider myself tougher than most. It's kind of stupid now that I have so many injuries, but I cultivated a badge of pride over not just being an actor who complained about everything. When one of the girls I worked with would stomp on set and say, "They don't pay me enough to do this shit," in front of the crew, ugh, I just wanted to let the ground swallow me up, because I was so embarrassed. I thought to myself, *Do you not have a concept of what the real world is like? Do you have no idea how hard it is out there?* I became obsessed with working extra hard to prove that I wasn't like the others. I should've worked harder to protect myself instead of proving myself to these men.

RR got meaner and meaner, and more and more manic. He wouldn't let me sleep more than three hours a night. When I was back in LA filming *Charmed*, one night I came home to my house, went into my bedroom, and screamed. There was a man in my bed and he was wearing a cowboy hat. Turns out it was a dummy. I assume RR put it there to scare me in case I came home with a man. I felt like it was just another move in what seemed to me to be a campaign of terror, the point of which was to destabilize me and have me fully under his control.

I started that movie at 112 pounds, and I went down to 97, and he would get mad at me every day for being too skinny, but I was so stressed out I couldn't eat; even water would come back up. You could see my ribs.

My only salvation during this time, once again, were my two Boston terriers, Bug and Fester, my two beautifully weird-looking gargoyles, with me on the set as usual. They did more for me than any human did. Sleeping with my dogs curled up around me was something even RR couldn't take away from me. These two beings were the only respite in the worst storm yet.

One night RR came into my trailer in a rage at midnight when we were about to take our "lunch"—because we would go to work at four thirty in the afternoon and get home around eight in the morning. In front of me he called Jessica Alba to ask her if she would step in and complete the film for me because it didn't look like I was going to be able to. He did it just to torture me; I didn't even know if she was really on the phone. I sat down and cried. I'm still not sure if she was on the line at all or if he was just pretending to speak to her.

He shocked me when he told me that what he did for a living was harder than what my brother did. My brother was fighting in Afghanistan at the time.

This was someone who got mad when his set lunch, his chicken breast, wasn't cooked just right. He was babied within an inch of his life. This was when I realized I hated RR. It had taken me this long and now I realized it was too late. I was trapped. Once again, the exhaustion made it hard to think straight. I couldn't see a way out.

One night, after he was finished raging at everybody on set and stomping away, with dark menacing eyes, he came into my trailer. He'd already fired me three times earlier that evening on the set, saying, "Don't bother to come back," and I was so freaked out, because if I got fired off this movie, I would never work again, ever. I knew the whole industry was watching, salivating, like a bunch of circling hyenas.

This time he declared he was going to fire me because I was secretly in love with Quentin, wanted to sleep with Quentin. That was as far from reality as it got. RR stormed out of my trailer, saying he'd be back with a lie detector machine to see if I really did love Quentin. Oh my God, as if.

I hyperventilated, as a full-on panic attack gripped me. I couldn't breathe. I called my then manager, a bigwig in the management game. It was midnight when I called, hysterically crying. I told him everything that had been going on, and I told him RR was coming back with a lie detector test and was going to hook me up to it. What I hoped my manager would have done was get me out of there immediately. But he either couldn't or wouldn't help me, and I was stuck. We got off the phone and I curled up into a ball and sobbed, trying to catch my breath. Once again, I was on my own and I was terrified. I knew that stress made people fail lie detectors, I'd seen that on TV. My heart rate was through the roof. An unsafe man was coming back at any minute to hook me up to a machine and verbally assault me. I furtively asked several people on the crew if they had a Valium. Luckily someone did. I swallowed the pill, praying it would work in time. I went back to the trailer and waited for him to return. I willed myself to be calm, breathing deeply. I felt nauseated.

Bang, bang, bang! My door got thrown open and there he stood holding a cruel-looking yellow machine with electrode-like wire things. I thought I was going to pass out from fear. My heart was racing so hard it hurt. He charged into the small space, his large frame filling the tiny tin box that was my trailer. I shrank back against the wall and started to cry. He told me to sit; I sat. I was pleading with him, tears running down my face. He attached the machine to me, turned it on, and started in, each question feeling like a bullet. I thought I was going to throw up, my stomach tight and clenched. I could barely breathe I was so afraid. The questions were all along the lines of, "Did you fuck Quentin?" It was laughable, but in the moment, the situation was no laughing matter. The crew probably thought we were having sex in there, but no, I was being terrorized. The lie detector said the answers were unclear. RR ripped the wire off me and stormed out of the trailer and back to the set. It was after midnight and I still had six more hours of filming to go. I didn't know what else to do, so I went back to the makeup trailer to get fixed up and get camera ready. My poor brain was scrambled, my nerves shot, and now I had to go act like I was the toughest chick around and save the world.

You know, if you say, "The director was a nightmare to me on the set," people will instantly say, "Maybe he was just trying to break you down to get in character." No, my character was the strongest woman alive. My character saves the world. This was just abuse by a man drunk on power with no one willing to stand up to him. When no one stands up to protect you, you stop expecting them to, it takes a toll. I started to feel like I was in a backward world, and I was losing my grip on sanity. My mental health was breaking down. I didn't know where to turn or who to turn to.

But I refused to let what was going on behind the scenes affect my work. If my child's name was going to be stolen for this character, I was going to be fiercely protective of my Cherry. Given the psychological trauma, insanity, lack of sleep, weight loss, day after day from all sides, and being totally alone, I'm proud of what I did. I'm really proud of the complete character I created. To be physically invincible, to be emotionally vulnerable and sexy, was not easy. If they gave out Oscars for pain and suffering, I'd have been a shoo-in.

Back on the set, there was a scene where I had to do a backbend over another actress. RR kept saying, "Higher, higher. You're not getting your arch high enough." To do a backbend in a four-inch high-heeled boot is already challenging, but with one of your legs set straight in a heavy weighted cast, it is pretty much impossible. I told him my arm hurt and I couldn't. He kept pushing me. I tried so hard to please him, my body the offering. RR pushed more, "Get your back higher." I used my last bit of strength and then *SNAP*. I felt my arm, like a cable in an elevator shaft, snap. Hot fire burned in my arm and brought tears to my eyes. I shoved them back down and finished the scene. I didn't know it then, but the snapping I felt was severe nerve damage, which would ultimately lead to my being paralyzed in my dominant arm. I thought I was going to faint from the pain.

Instead of being taken to a hospital for my arm, I was taken outside to film the next scene, that of my boyfriend dying after being murdered. Yes, RR knew about my ex-boyfriend Brett's murder and had written a scene that would put me back into the mental space I'd been in when he died. "Art" was imitating life. Even though my boyfriend character was dying because of a pseudo-zombie attack, I still had to go through the very real emotions of my boyfriend dying in my arms.

I was in so much physical pain, my arm was on fire, but seeing my pretend guy's body under me, covered in (fake) blood, and to go through the emotions of said boyfriend dying . . . it was a mind/soul fuck, let me tell you. As an actor, you trick and abuse your emotions, bringing pain up only to have to shove it back down when you hear, "CUT!" I believe acting, dramatic acting, is a form of self-abuse. Take after take of pulling up dark emotions only to seal them up when the camera stops rolling lest you make anyone on the crew uncomfortable. I cried until RR said cut for the final time. I was depleted, in pain, and exhausted. There was to be no rest because a core group of us had to go to Comic Con, the biggest comic and film convention. I had to go make a personal appearance and wow the "fans." In other words, be "on."

At Comic Con, with my elbow resting on a giant ice bag, I signed hundreds of posters and sat smiling vacantly. My arm was on fire. I would take crying breaks in the bathroom, but—and this is so emblematic of my "be a good little girl" Hollywood training—when I had to cry, I stood up and made my face go horizontal so my tears dropped on the floor and not down my face. Couldn't ruin my makeup.

Somehow I completed the film. My arm got more and more painful. I was losing complete control of my arm. I looked like a skeleton. I was a very physical person: fight training, boxing, stunts. Now I couldn't hold a pencil or a fork with that hand without shooting pain that is unknowable unless you've suffered from nerve damage—it's like an electrical firestorm inside your own body.

I told RR I needed to get surgery. He told me not to claim it on his company's insurance—that I had to pay for it out of my own pocket. Over the next couple of years, I had two surgeries on my wrist, and two on my elbow, which wound up costing me thousands upon thousands of dollars, but more importantly, it cost me my ability to ever be as physical as I was. I am still damaged, and I still live with a lot of pain.

The next blow was the worst of all. RR sold our film to my Monster. Yes, the Monster's studio was going to release our movie. I can't tell you what it was like to be sold into the hands of the man who had assaulted me and scarred me for life. I had to do press events with the Monster and see photos of us together, his big fat paw pulling me in to his body. In the end the film was a box office flop, I think largely because they promoted it horribly, but to tell you the truth, I was happy it failed. I was happy these men wouldn't be making money off me.

Everything was upside down, backward, and insane. Things that are completely outrageous had become the norm. That's true of abusive relationships. It's true for the cult. And it's true for the cult of Hollywood.

Being abused by the Hollywood cult breaks you down, as anybody who's been abused knows. They break you down softly at first, telling you things like "I accept you" in a gentle voice. They whisper words of love and devotion that nobody will ever love you like they love you. And this is exactly the same thing a cult leader tells you. And that is what makes it so dangerous. While my arm was still in a cast, I decided it was a good time to get a sinus surgery I'd put off. I'd had terrible sinus infections since childhood and had horrible trouble breathing out of my right nostril for years. When I went to the ear, nose, and throat surgeon, something went horribly wrong. Somehow going from above my sinus cavity and underneath my right eye, a puncture was made through my skin. I still don't understand. I was told by the doctor to go home and that under eyes don't scar. I kept waiting for the hole to heal, but it didn't. After a week, I went to a plastic surgeon. He immediately performed a surgery to make the hole a thin line.

I had to get reconstructive surgery on that eye, and then, because the procedure pinched the eye a tiny bit higher, I had to get my other eye slightly done to match it, by one millimeter. I told my publicists what had happened and they said to say it was a car accident. Looking back, I don't know why it mattered, but I took that advice. And so when I was asked by the press, that became the party line. Early on after the real accident, there was a paparazzi shot taken of me where there was a ropy scar under my eye. It started the plastic surgery rumors. They put me in *Star* magazine and "Knifestyles of the Rich and

Famous." They had a split picture of what I used to look like and what I looked like now. Perez Hilton would put me on his site with, "How old is she? 65? 75? 85?"

For the next year I visited a doctor four times a week, a needle in each side of my ropy scar, so they could break up the scar tissue. Lasers burned my flesh to flatten the knotted skin. The trauma was real. The pain was real. The harassment was real.

One night I crumbled. I was huddled over in a bathtub and I let go. I wailed, I cried. I just wanted it to stop. But then I got out of that bathtub, dried myself off, and marched on.

I was scared, sad and broken, and lost. A breaking point was coming.

Ten months after we finished filming on *Planet Terror*, I found out I'd be on the cover of *Rolling Stone*, with Rosario Dawson. Even I got excited. A cover shoot for *Rolling Stone* is the Holy Grail for an actress. It's a cultural landmark. In reality it was the tipping point that finally snapped me out of the Hollywood brainwashing and got me on my path to being whole.

I showed up, said hi to the photographer, who had shot me before for the *New York Times*, when I noticed something was missing. "Where's the wardrobe?"

He looked at me sheepishly. "Oh, there isn't one, but we have a great idea! You'll be butt to butt with Rosario, with a belt made of ammo around you! How HOT is that? Isn't that genius??"

Another man helping to manifest misogyny. I sucked it up, put on that dumb belt, and stood on an apple crate so my butt would perfectly line up with Rosario's (she is taller than me). They tried to re-create my Cherry Darling look, with the extra-large hair and fake tanner, but no one could ever get me as fake tan as I needed to be to get through this bullshit.

As I stood there, pursing my lips, touching butts with Rosario, I had a realization.

I was being photographed by a gay male who was imagining me as what he thinks a straight man wants to fuck, and he was doing so on behalf of a director, a straight male, who was interpreting me with his little boy brain on behalf of the studio, also male, who were interpreting me based on who they want to sell tickets to, which is this invisible horde of boys and men. The male gaze is real, ladies and gentlemen, and it is deep.

I knew I had to do something about what I'd let my life become. Something kept nagging at me, the wrongness of my life. It felt like an insistent alarm bell in my head, but I couldn't quite figure out what that alarm bell was trying to tell me. Something needed to change, but I didn't know how or what.

Then I saw my *Rolling Stone* cover on the newsstand. I backed away from it slowly, my eyes wide. The photograph looked nothing like me. This was a moment, meant to mark the epitome of success as an actress, and I couldn't even recognize my face. The alarm started ringing in my head. *Wake up, Rose, wake up, Rose.*

I stared at the magazine, wondering, *Who is that person? What have they done to you? This is what it's come to?* All these years of being sexualized, it was crystalized in this one image. There is a difference between being sexual and sexy in and of itself. There is nothing wrong with that. Sexualized is when others have done it to you for their benefit. I had a hand in it. I didn't say no, but like so many women in cults, I didn't know I could. I had lost my voice. I had lost myself. But my *self* was desperately trying to wake me up.

PART TWO

ASHES TO ASHES

Looking back, if there were one thing I could change, it's this: I wish I had known I was an artist a lot sooner. I would have started directing a lot sooner. I would have started singing and writing a lot sooner. I wish someone had just said one thing to me that made it click, but, you see, there were really no examples to look to. I didn't see where I could become what I could be, because it didn't exist. And the cult doesn't want you to know your value and it certainly doesn't want you to know you are free.

So I started with the small stuff. I went back to the business of getting healthy physically. Then I took time to heal mentally. A long road ensued.

I have something called cyclothymia, which means I am

prone to depressions, which I now deal with by taking a mood stabilizer. There is no shame in this. The depression really rocketed off after my assault and became something impossible to manage on my own.

I'm only talking about this to help others who may also suffer from the mean blues, as I call them. What I know now is that you should not be afraid to seek medical help for mood problems. Try one thing, but if that doesn't work, try another. It is a needle in a haystack finding the right protocol, but it can be life changing. It's like the blinds raising on your windows and you can not only see, but also feel the sun in your life.

I began the process of putting my life back together and turning my back on Hollywood. During this period of healing, I was able to work on my relationship with my father. He had spent some of the previous years reaching out and repairing relationships with my brothers, sisters, and me. He wasn't perfect at it, and he and I still had a relationship where we were more comfortable talking about anything but emotions. He and I bonded the most discussing politics, the war in the Middle East, Sufis versus the Sunnis, anything but our feelings. We still danced around the past, but we could talk well enough about the now. We laughed a lot, too. He was one of those people who made you feel like a million bucks if you could make him laugh. His mental illness had chilled out at this point and made him much easier to be around.

It took a lot to love him and have my heart unfreeze, but during my later Hollywood period, he showed me kindness and was one of the few in the world who did. He hated Hollywood men and how they had hurt me. It made me feel good that he cared. We went from a family that was very fractured and damaged to having him really be back at the center, not as the demon, but as the bright beautiful funny light that we all gravitated around. He had somehow raised eight kids in his own haphazard, fucked-up way, at times magical, at times hell.

Then he got sick. I was with him at a hospital in LA when the doctor shook his head and said, "I'm so sorry. You have pulmonary fibrosis." I had no idea what that meant, but quickly learned that pulmonary fibrosis is a death sentence. My dad was given a maximum three years to live. Every time you breathe, your lungs rip apart and scar, and the scar tissue builds up and overtakes the airways. So every breath you take is one step closer to choking all the air out of you.

It's a horrible, horrible disease and a terrible way to die. It kills more people than breast cancer each year, and yet people have barely heard of it. That's because for a long time, it was affecting older people, but now it's dropping younger and younger and younger. It's idiopathic, which means they have no idea what causes it. For my father, I think it was from the airbrushes: years of inhaling tiny particles of paint.

He was fifty-eight when he was diagnosed. It was wild because he was Mister Healthy. Never owned a microwave. Never ate processed sugar. Was on the Mediterranean diet way before it was a diet, ate super organic. He put our cereal in paper bags to avoid the plastic bags.

Every one of us kids—a few of us who'd plotted to kill him at one point or another—were all lying with him on his hospital bed when we made the choice to end his life by disconnecting him from life support. A beautifully complex star was dead.

After my dad died, it was like nothing mattered anymore. I went into this haze and all I did was watch classic films and hang out with my dogs. I began the slow climb from a beaten-down individual to being a strong-ass woman, to being somebody who's not afraid to be herself, to know my own worth. Slowly I realized the healing process was beginning to evolve into a transformation. Mine.

During the next few years, I worked hard on myself, slowly fusing the many characters I played into one person. A whole person. Me. I was a girl interrupted and when I went back to my regularly scheduled life, I didn't really know where to be or what to do. I did a few jobs here and there, but in my mind I was getting out of the business. I was still unraveling a lot of the fear and trauma my life had brought on me, healing physically and processing my father's death. My thirties were kind of a wash. They were just spent in a haze. I wish I got a do over. Maybe that's what I'm doing now.

Raised as an outsider, then a runaway, then a captive, then a starlet, then a celebrity, it was finally my turn to be me, just a member of society. A fully fledged human. Fully myself. To get to be me from the beginning of my day until the end of my day. To feel what *I* want to feel and not what I'm *scripted* to feel. It had been a long time coming.

I am no longer willing to be a part of something I wouldn't want my mind and soul to consume.

My friend Joshua Miller (the one I corrupted at thirteen), along with his writing partner, M. A. Fortin, came to me during this period and simply said, "It's time." Meaning it was time for me to direct my own movie. They wrote a brilliant script for a short film, *Dawn*. As the director, I set out to make a minimasterpiece. I wanted to explore what happens to girls when we send them out into society trained to be polite and little else. *Dawn* is set in 1961, but I knew it would speak to current audiences. On the set of *Dawn*, I kept waiting for my nerves to kick in like how they did when I was acting, but the nerves never came. I was the captain of the ship and I damn well knew how to sail. After working on sets for over fifty-seven thousand hours, I know how to run a crew and set, but more than that, I know what I want to say in a medium I love. Directing for me is as natural as breathing. I truly found something I loved. Finally. It was fitting that the first time I returned to Sundance since my assault was as a director with my *own* film nominated in 2014. *Dawn* was nominated for Grand Jury Prize, a huge honor. Since then *Dawn* has been the little movie that could. Screening at Sundance London and Hong Kong, it was toured around the United States in theaters, screened at the Lincoln Center, and is now on the Criterion Films/TCM platform, FilmStruck. To say I'm proud of everyone involved is an understatement. After directing my first piece, my confidence as an artist and independent thinker grew exponentially. It was time for me to use my voice. But how?

PHOENIX RISE

It was Ashton Kutcher, of all people, who got me involved with social media. I have never met him. He said in some interview that Twitter is the only voice an actor can actually have.

Something clicked for me. I started slowly on Twitter, just easing my way into it and just saying whatever, because isn't that the point? It amused me. Then one night I was sent a script by my agent at Innovative, a second-tier agency in Los Angeles. I was using them for voice-over work only, but somebody there got the brilliant idea to send me an Adam Sandler script. The role in question was a former model who's obsessed with Adam Sandler and stalking him. Imagine me, stalking Adam Sandler. Hahahaha. Attached to the script was a wildly demeaning cover letter and helpful instructions for the audition. I took a photo and posted it on Instagram and Twitter. I went to bed, not thinking anything of it.

> **Notes:** -Please make sure to read the attached script before coming in so you understand the context of the scenes.
> -Wardrobe Note: Black (or dark) form fitting tank that shows off cleavage (push up bras encouraged). And form fitting leggings or jeans. Nothing white.

I woke up the next morning, checked my phone, and realized my tweet was going viral. My heart started to race. All internet hell broke loose. It was picked up by media in India, China, everywhere. My second, deeper wake-up came when I started reading my Twitter followers' comments and realized people were rightfully aghast. What's sad is how normal this was to me. I'd thought it was just an eye roll, a bon mot. One tiny example of infinite indignities suffered in Hollywood by women. But the reaction made me realize it was more than an eye roll. That's still on me, to pull out the threads of this misogyny, sexism, and the brainwashing of Hollywood and media, to free my own mind.

I realized it was time I finally start some real conversations with the public. I would no longer be silenced. I had known my activism was going to come out publicly at some point; I'd been an activist for gay rights for years, just not an activist for women specifically. I would never have expected Adam Sandler to be the big reveal. But in retrospect it makes sense. I mean, in what world does Adam Sandler—representing the average American male, naturally—end up with Kate Beckinsale, Salma Hayek, or Jessica Biel? The "silly" movies he's known for are more dangerous than they appear, because they propagate a sense of entitlement among the millions of dudes who are his primary audience. The message is that they, like he, in his stained T-shirt and sweatpants, and by virtue of possessing a penis, having no discernible mental assets, have a right to an accomplished, amazing woman like a Salma Hayek. And that is communicated not just by the shitty script saying so, but by the casting itself. It's absurd. The messaging given to men by the media is about entitlement and ownership, following privileged boys' club rules. And they are the voice of society. There is definitely a parallel between this mind-set and violence and sexual harassment against women.

I was in New York during the Twitter hurricane that followed. There was a screening of my film *Dawn* at the Lincoln Center, a huge honor, and I went right from there to film *Watch What Happens Live*, a show on Bravo hosted by Andy Cohen.

I walked off set when I was done, thinking what a waste of my life that was, when I got an email from my agent, saying they were firing me.

For about ten seconds I was gripped with a paralyzing fear. I thought, *Oh shit, it's happening again. I'm going to be blacklisted all over again*. I could feel myself shaking. But then I realized—

The agency honchos thought they could fire me and it would stay private. Big mistake. They thought I was going to protect them, because like I said, Hollywood works like a Mafia.

I started to think on it. Why should I protect these purveyors of brain smut with my silence? They undersold me, undervalued me, and underwhelmed me. They have never protected me or legions of other girls, women, and young boys. They've done nothing to protect the sensibility of society, and they've done nothing I can see that's moving society forward, at least not lately.

Hollywood operates under a veil of secrecy, but I never said I'd be a secret holder. If you don't keep their secrets: *Tsk, tsk, tsk, we know what you are. You're a loose cannon. You're never going to get ahead. You, little girl, will be punished.* How about this: How about you don't do disgusting things that you need to keep secret? How about this: How about you treat women who work with you fairly and humanely, the same way you treat the guys? How about you put yourself in their shoes and see what it's like to experience their lives? The only way to change things is to lift the blanket and shine a light on the darkness. My goal is to expose Hollywood for what it really is. I have a million more stories I could tell in this book, but I don't have the energy to document all of the dickery. By using a few of my stories as descriptors, I hope it will have a domino effect. More women will rise up, take their power, and say "no more." And men will stand as allies.

So I tweeted again about being fired and punished for speaking up.

It's a wild sensation when something I write or do goes viral. All I have is a phone in my hand, but it's a conduit to so much energy. It's the way to tap into the world outside and inside other people's thoughts. It's been a really useful tool for me to speak for myself, to speak out, and to speak up for others, but it's still a uniquely strange experience when your phone feels like holding a live wire in your hand.

I finally realized how important it was for me as an artist to claim my strength and my power and my worth. Because there's something inside of us that they can't take away, no matter how much they try. I'm electing to try and change societal norms, so it occurred to me that I'm a politician of sorts. Except I can't be voted out of office. If you're an artist, no one can take that from you.

Soon enough, *Good Morning America* came calling. It was a great interview. I was passionate and the questions were respectful. You generally only see a woman be intense on-screen when she's being two-timed by a man or something. You rarely see intense women being interviewed. We women have historically been trained to be pleasant at all times. We need to stop that and be authentic; justifiable anger is a part of that. Being angry is okay; no one is going to die if we women let our anger out in healthy ways. It's more than okay for women and girls to have emotion that is their own and not connected to what a man does or society says.

Just as the Adam Sandler of it all was dying down, something else happened that I knew I could use as an example for change needed. In June 2016, the new head film critic of *Variety* magazine, Owen Gleiberman, published a nasty piece shaming the actress Renée Zellweger for, in his estimation, appearing different than she had the last time he'd seen her, implying she had "had work done," and asking if she is still the same actress if she "no longer looks like herself." I decided to push back at this shameful excuse for journalism by penning a public letter to its writer and breaking it down so others could see and understand that it's not really about Renée Zellweger, it's about all of us women:

> Owen Gleiberman, this is not a counterpoint. There is no counterpoint, there is no defense for the indefensible.
>
> Renee Zellweger is a human being, with feelings, with a life, with love and with triumphs and struggles, just like the rest of us. How dare you use her as a punching bag in your mistaken attempt to make a mark at your new job. How dare you bully a woman who has done nothing but try to entertain people like you. Her crime, according to you, is growing older in a way you don't approve of. Who are you to approve of anything? What you are doing is vile, damaging, stupid and cruel. It also reeks of status quo white-male privilege. So assured are you in your place in the firmament that is Hollywood, you felt it was OK to do this. And your editors at *Variety* felt this was more than OK to run.
>
> You are an active endorser of what is tantamount to harassment and abuse of actresses and women. I speak as someone who was abused by Hollywood and by people

like you in the media, but I'm a different breed, one they didn't count on. I refuse and reject this bullshit on behalf of those who feel they can't speak. I am someone who was forced by a studio to go on *Howard Stern*, where he asked me to show him my labia while my grinning male and female publicists stood to the side and did nothing to protect me. I am someone who has withstood death threats from fan boys, had fat sites devoted to me. I've withstood harassment on a level you can't comprehend, Owen.

I was so confused by the heaping tons of abuse, I actually forgot what I looked like. Which is awesome because I rose up from some serious ashes to finally have my say. Here's some truth: Men like you and the women who sit idly by and say nothing should know that aiding and abetting is a moral crime, and if it were punished in Hollywood, most of you would be in some form of jail.

Any studio that Renee Zellweger has made money for, any co-star she's supported or anyone who takes a percentage of her income should be doing what's right: They should be calling this harassment out.

As a woman who has been bullied for years by a vicious pack of lower beings, I can relate to this. Many are probably silent because they do not wish for the proverbial pen to be pointed at them; I say point away. Short of killing me, you can't possibly do more than what was done to me in my tenure as an actress. I don't care if you're afraid. Be brave. Do what's right, for once. I loathe fear. And this town is built on fear. Fear was instilled in me by the men and women of this town, just as I'm sure it was instilled in Ms. Zellweger. Fear of being blacklisted, fear of being branded difficult, fear of . . . fear of . . . fear of.

Well, guess what, Owen? I am not afraid of you or anyone. It is a small, small, myopic, self-fellating town that loves to love itself. I am here to ask you all to put the mirror down and look out at society, because whether you're aware of it or not, you too are part of society, and by retreating to the standard go-to—silence—you are hurting all of us. Look at what you're doing and where you bear responsibility and culpability. Who are you all protecting and why? Who are you helping and why? . . .

Owen, the last line in your article, "I hope it turns out to be a movie about a gloriously ordinary person rather than someone who looks like she no longer wants to be who she is" is quite the mind fuck.

Guess what? It is time to stop fucking with women's minds.

Do you know what my interests are, Owen?

My interests are bigger than pondering a stranger's face. My interest is destroying the status quo. My interest as a card-carrying member of society is to STOP the brainwashing Hollywood and the media have for too long gotten away with. The brainwashing that you have long been a friend to and a supporter of.

Let's talk about Hollywood writers: Joan Didion, John Fante, Raymond Chandler, Robert Towne, Dorothy Parker, John Gregory Dunne, Preston Sturges, I.A.L. Diamond, Pauline Kael and Billy Wilder. These were *writers* on Hollywood.

You, Owen Gleiberman, are not they.

You are simply a bully on semigloss paper.

Renée Zellweger never reached out, but I didn't really expect her to. I didn't really do it for her, I did it for all women who get affected by this strain of bullying and shaming, myself included. People in the industry do not go up against *Variety* or *Hollywood Reporter*, ever. They want a good review. The problem with these dailies is that it's often the only news that Hollywood cult members consume. And it is white industry dudes talking to white industry dudes. Echo chamber. The way they talk about women is a big fat yuck. They reinforce the fucked-up ideas so many in Hollywood have. My reality is that I deal with worldwide press. Two little industry papers aren't really of interest to me. Not until they change their ways. Also, it turns out that for years *Variety* was on the Monster's side, doing his bidding.

A few months after that *Variety* piece ran, I noticed billboards everywhere for a new X-Men sequel. Twentieth Century

Fox was the studio releasing and marketing this film, squarely aimed at tweens and teens. Guess what the ad campaign was? Jennifer Lawrence, America's latest sweetheart (and a respected Oscar-winning actress), being strangled by a big stone-faced "man." The tagline was "Only the strong survive." I had read that many women were furious about this campaign, but that Fox was turning a deaf ear. I am assuming no one in Lawrence's camp realized how wrong this marketing campaign was. Whoa. Red flag. This campaign to most in Hollywood looked completely A-okay. Fuck that. Once again, I decided to write an open letter. I felt an obligation to be a voice for women who have suffered from violence. I wrote:

> There is a major problem when the men and women at 20th Century Fox think casual violence against women is the way to market a film. There is no context in the ad, just a woman getting strangled. The fact that no one flagged this is offensive and frankly, stupid. The geniuses behind this, and I use that term lightly, need to take a long hard look at the mirror and see how they are contributing to society. Imagine if it were a black man being strangled by a white man, or a gay male being strangled by a hetero? The outcry would be enormous. So let's right this wrong. 20th Century Fox, since you can't manage to put any women directors on your slate for the next two years, how about you at least replace your ad?

> I'll close with a text my friend sent, a conversation with his daughter. . . . Her words: "Dad, why is that monster man committing violence against a woman?" This from a 9-year-old. If she can see it, why can't Fox?

In a statement, Fox apologized for the billboard: "In our enthusiasm to show the villainy of the character Apocalypse we didn't immediately recognize the upsetting connotation of this image in print form. Once we realized how insensitive it was, we quickly took steps to remove those materials. We apologize for our actions and would never condone violence against women." To Fox's credit they took down the campaign. That said, they wouldn't have had I not spoken up. And they still haven't hired any female directors. Once again, it's an all-male voice and perspective being mansplained onto the world.

I faced blowback and harassment from the "I live in Mommy's basement with my laptop" boys who could not seem to grasp that the image is a major reinforcement of violence against women. "She's blue! She's not real!" Like I wrote in my protest piece, "if a 9-year-old can see it, why can't you?" Willful stupidity gets so exhausting to deal with, sometimes I just want to scream. I know I am not alone in this, but I am out in front and therefore prey to abuse. But it is for the greater good, even for the good of those whiny basement dwellers.

Then came election year. In October 2016, as women came forward with their experiences of being assaulted by Trump, there was the usual pushback from rape apologists, which sparked an onslaught of comments about #whywomendontreport—particularly when the assaulter is a prominent individual. I tweeted:

And finally, I wrote this:

To the women and men in the entertainment industry who know exactly whom and what I am talking about, I say be brave. Do not work with those you know to be offenders or you are no better than they. Take a stand. You are culpable for your actions. Stop rewarding sociopaths. Every time you sanction abhorrent behavior, you are aiding and abetting a crime, that makes you no better than the criminal. How many more stories do you have to hear before you do the right thing and stop rewarding men that are predators? Why are you so cowardly that you would take the softer, easier way out? I can tell you this, your soul is a blighted one if you do so. Your personal legacy, the very fabric of your being, is at stake, so fight for it. I know you have it in you to be better. I know you have it in you to break free from the bonds of secrecy. So do it.

I've figured out a rhythm of engaging with people, using Twitter as a vehicle for it, and Facebook to an extent. I have learned to use social media as a conversation starter. I look for ways to try to join worlds, to build that bridge to reframe status quo ideas and ideology. I don't always get it right, but my intentions are pure.

I don't use the word *fans*; that's not how I interact with the public. It's not a fan and celebrity: we're all humans, in it together. I consider my supporters to be cothinkers.

Above all, I know I speak to people like me: the disenfranchised, the hurt, the lost, the lonely, the brave ones who choose to live their lives differently, who choose to see things differently, who choose to function in a society that doesn't want them in their own way, on their own terms. These are the people in my tribe; if they are not in yours, I suggest you get to know some. Your life will be richer for it.

What if I'm right? What if we could live a better, freer, stronger life? Stand up and do what's right. What if people had a conversation that everyone is so afraid to have?

That's what the #RoseArmy is about. It's a hashtag used on my social media by those who identify as freethinkers. We are an army of thought. We are a group of like-minded individuals who see things differently, who live differently.

We stand for freedom of spirit and mind. We stand for dissent. We stand for equality and the right to have our lives be unencumbered by proscribed thoughts and traditional mores.

I highly recommend independent thought and critical thinking.

I also think daydreaming, if you want to call it that, is essential to development, just to space out on a couch and think through life and break it down, break your situation down instead of reacting. By reflecting and giving yourself time with the TV and internet and music off, just to hear the sound of your own mind, is an incredibly important thing to do. I highly advocate it. Be brave. Be brave, look inside, consider your positions on things, ask why you feel that way, go down into that rabbit hole of thought. Get meta. Get deep. And for those who do, go even deeper.

Most of all, I recommend creativity and taking a creative approach to every aspect of life. During the time I wrote this book, I also recorded an album, *Planet 9*. I decided if I can't go to another planet directly, I will just have to create my own interplanetary realm. My album is an experience that will take you on a trip through time and space. Using my own voice and words to create and elicit emotion is a powerful thing and I am very, very proud of my album. It's ethereal and cerebral and propulsive and makes you feel.

Lyrics to the chorus of "RM486" are "only here to paint colors on the sun, only here to see the fires run" because that is all we are here to do. We are here to be and bring magic.

Artistic thought is something I believe exists in every single human on the planet. It just gets beaten out of too many of us instead of being valued and nurtured. Creativity is freethinking. Innovation comes from creativity. When I say I'm a proponent for thought, I mean creative thought. If you look at how society works and you look at how it can be improved, then you're looking at something creatively.

Art feeds thought and art is everywhere. It's not just inside of museums, which can be intimidating and leave people unmoved. I'm in Berlin as I write this, staring out my window at a hospital. It's a beautiful piece of accidental art. Each of the windows in this hospital is a different shade of amber and green, so it looks like a gorgeous art installation. Art is everywhere, indeed. I've started to even show people my photography on my @rosemcgowanarts Instagram page and am planning a photo exhibit of my works. I like to photograph people not necessarily even because of what they look like but because of how I see their architecture, the shape of their face cutting into air. I see the lines of it, and that's what I want to capture. Art is in the angles in the corners of your room. It's in how the light kicks off the water. It's everywhere, you just have to look for it.

So many businesspeople I meet wave their hands while saying adamantly, "I'm not on the creative side." And I think, *Wow, you've been really messed with, haven't you? Why wouldn't you want to be on the creative side of everything?* When I hear someone say they are not creative, I immediately think, *You poor thing, they got to you. At what point was your creativity stolen from you? How old were you when they homogenized your mind?* Because that is what noncreative thinking is: homogenized thought. And it keeps us all so separate.

Why on earth would any businessperson not be creative? How are you supposed to be great at your job or life? How are you supposed to be passionate about something if you don't look at it creatively, if you don't see and try unique ways of doing things? I'm sure you've had situations in your life that have called for you to be creative so as to figure it out. Boom! Guess what that means? You are a creative. You are an artist in your own life.

I think it's bizarre and tragic how society pushes us to say we are what we are because of the job we do. The question "What do you do?" really means what do they pay you for, as if that's your defining characteristic. Everything else is a "hobby." But those are also things you are and do. Just because you don't make money at it doesn't mean you're not doing it. It's as valid as going to the office, maybe more so.

By removing labels that people put on us, and that we put on ourselves, we can have a much richer life, so much more adventurous, so much more fun.

How do we remove these labels? I started by writing down my beliefs about myself and traced back to whom I got them from. I started to think deeply and look at the beliefs holding me down, knowing if I got to the root cause, I could work to be free. Because guess what? By sticking to old belief systems, I was most definitely not benefiting my own spirit. By writing these old beliefs down, I was free to at least think differently about myself and carve out a different future. One based on my true strengths. One based on how *I* truly saw myself, not how I was seen.

CULT OF THOUGHT

Let's discuss entertainment as male propaganda, shall we? Questions: Can you imagine if man's history was only depicted, shown, and viewed through a woman's perspective? Do you think there has ever been a film where only women were hired to tell a "man's story"? The answers are no. Rarely is this commented on or spoken of, because once again, we are all too used to table scraps as women. Women in the world have been interpreted and sold back to us in a dangerous way. We are mirrored back to ourselves almost exclusively through a man's idea of what we are; it is gender appropriation in the extreme. Fact: the Director's Guild of America, the union representing working directors in Hollywood, is 96 percent male and it has been since 1946. That means for your entire lives you have been fed a steady diet of largely male "thought" and bias about what women are and can be. Why is 96 percent of our information, entertainment, and philosophy coming from men? It's because of systemic misogyny. It's because of zero government oversight. Film and television is a white male's playground and through their narrow vision comes the manipulated mirror you're largely forced to see yourself in. If that's not male washing, I don't know what it is. In 2016 only 23 percent of speaking roles on-screen were women, the majority in horror films, and we all know how they get treated. If you don't think this is a big deal, you are wrong; it is a very big deal. This is how we form our notions about ourselves and others, and it poisons us in ways we are not always cognizant of. As of now there is no answer for this injustice; all I can tell you is to watch what you consume. Start noticing the stereotypes and the clichés. Start rejecting them. Start complaining. Tweet at the directors, studios, whatever company is behind them. But most of all, demand more. Your mind is at stake.

A not so strange thing happened while I was writing the end of my book. A producer and writer of a very famous cable show that grossly used women as objects and props was texting me things that were wildly inappropriate. I had never engaged with him in this way. This man who sold his perspective of female ownership to millions of viewers for years. The exchange was filled with him being manipulative, pushy, and ultimately, sexually inappropriate. I rolled my eyes at this creep, but then I thought about his messages a little deeper. I realized that if he felt okay saying this stuff to me (!), me who's obviously someone who so often calls bullshit on offensive and stereotypical bullshit male behavior, well, what is he saying and doing to other women? And even worse, he continues to spread his disregard for women to a global audience, by having them act out what he thinks women are. And so many of you have just eaten it up thinking it's harmless, mindless entertainment. But it's not. Everything you consume counts. It forms you, it matters. Know what you are watching so you can reject it. If Hollywood can't change, it deserves to fail.

I'm so happy to push back at the machine that for so long has affected our minds with its narrow perspective. Because when you push back at Hollywood, you are pushing back at the stereotypes the industry peddles. And here's the thing—cinema is dying. Do you know why men in Hollywood have run out of ideas? They've run out of ideas because they've fucked themselves. They've marginalized themselves by clinging to misogynistic ways and outdated ideas. And their steadfast dedication to keeping white male power in place is costing them billions, that's how entrenched they are in their rightness. They aren't even good businessmen, but they've got a hold of our minds.

A centuries-long smear campaign against women has gone on for far too long. I call time. Tell me the truth, does the woman on-screen remind you of your mom or your sister or your aunt, or the little girl down the street? Not just in looks; I mean in action and thought. Women have interests, hopes, dreams, aspirations. Beyond being fuckable.

The fact that the vast majority of men in Hollywood see women in such an antiquated way has boxed us in societally. I would know, I was the one helping to do it to you. So many of my interviews with their insulting, condescending questions from the male-run media were about shaming me for being a woman who didn't fit into their narrow mold. The interview equivalent of saying a girl deserves to get raped if she wears a short skirt. Don't make your Madonna/Whore complex mine or our issue. Grow up. Think deeper.

In my acting career, I tried very hard to play strong roles that made an impact. No interviewer ever asked me about my craft, my methods—they don't often ask actresses that, especially not the ones who have to wear short skirts. Too many interviewers, male and female, approach female subjects with a consistent agenda: to marginalize, sexualize, ridicule. I've been told my whole career I'm too smart "for an actress." No, I'm smart, period. They did that to box me in and keep me in line. We have to recognize this gaslighting and push back.

Hollywood maintains this fiction that they're selling tickets to a young male audience. This is to justify the unending onslaught of objectified females. But newsflash, these young men are downloading your product illegally. They're not buying tickets. It's the women and the girls that Hollywood should be courting. That's who drives ticket sales; it's been proven over and over statistically. Women have been excluded from being directors, writers, cinematographers, tech executives, engineers, scientists, philosophers, bankers, and artists; we need to take whatever it is we want to be. Society will continue to be messed up until we start telling different stories and thinking deeper about our responsibility to the public. People go in droves when they see themselves on-screen. And guess what? We are not seeing ourselves and we are tired of it.

There's something particularly appalling in how media and Hollywood have gotten into lockstep with each other, operating hand in hand to dumb down the populace. If you don't read—42 percent of US college graduates never read another book in their life after they graduate—entertainment and online media are where your thoughts are being formed. Those media products don't provide you with a mirror of your life, do they? Do you see yourself on that screen? Probably not, so why are these men in charge of the mirror in your mind? When people finally understand where 99 percent of their media is coming from, I hope they'll switch off or at least be aware of what they are seeing, choose wisely, and start harassing these studios until they fix their ways. Twentieth Century Fox studios has no female director on the slate for the next three years. That means it's almost all white male content this multicolored world is getting. No wonder it sucks.

Earlier in this book, I told you about the classic film star Frances Farmer, the one who had her brain shocked. I was using an extreme example. But guess what? Even more tragic is that the same fate is befalling women today, as Hollywood continues to strip girls of their dignity. And if they're doing it here, they're doing it to you, too. I was recently in Miami and I met a beautiful young actress with haunted eyes, her hair professionally blown out with three curls at the ends, the de facto LA hairstyle that they stick on every young woman. She told me that when she was seventeen, a few years earlier, she was guest starring on a series with an actor who was known for his sex addiction. They called her mother to ask permission to show her butt in a shot, and the mother gave permission, but when the girl got to the set, they said actually, for this scene you'll be walking through the crowd, naked. Her mother wasn't on the set. The young woman didn't know what to do, so she did the scene. She walked naked through the crowd of extras ogling her and at the end of it, the male star got on his knees and went down on her, on camera. A seventeen-year-old girl. It was filmed. It was aired. And it was consumed by the public, maybe even you.

Everybody on that set knew the guy was a sex addict. The producers, the executives, the distribution agents. Everybody knows, but nobody does anything. These are weak fucking humans. I despise them. I despise them for still shrugging and reasoning: *it's just a girl.*

I despise Bill Cosby for being one of America's most prolific sexual predators. Judging by the number of women who came forward, you can only imagine how many didn't come forward because they either died or couldn't bring themselves to go through that media hell, being written about and talked about by willfully ignorant men (and some ignoramus women) who have no idea what it's like to be a victim. There's a whole machine set up behind the grotesque appetites of a star like that: men who scout for the girls, men who bring the girls in, men who hush the girls up afterward. That makes it a supply chain. It was a whole cottage industry. It's not just one person going to a bar and drugging somebody. It's systematic and it's not an accident. Agents, managers, entertainment lawyers, assistants, executives, unions . . . all had a hand.

Sometimes we don't even know we've been raped. We chalk it up to being a "sexual experience" because we don't know better. When I was a teenager in Seattle, there was a store where I liked to shop in Capitol Hill—more accurately, where I liked to try on clothes, because I couldn't afford to buy them. The place was on Broadway, a street made famous by Sir Mix-a-Lot in his classic "Posse on Broadway." The clothes were all black and so cool. I had browsed this store before, and the manager told me he'd give me a 20 percent discount. At this time I was being given one dollar to eat lunch, so that 20 percent wasn't going to do much good. But the manager seemed cool; he didn't treat me like a child although he for sure knew I was one. I decided to go to the store the next day to try on clothes and wish I could buy them. When I was in the dressing room, the manager came in. He was silent. I pulled my shirt across my chest and shrank back against the wall. I didn't understand what was going on, I hadn't asked for any other clothing. He pulled his pants down and advanced on me. I'd never seen an erect penis before. It was big and veiny and scary. I had no frame of reference for what was happening. He pulled down my shirt and put his veiny penis between my breasts and used them to get off on my chest. I remember detaching, floating away, like I was on the ceiling looking down. Praying it would stop and I could leave. I was in shock, trying to figure out what to do, when the man's wife came into the dressing room just as her husband was zipping up his pants. She shrieked, throwing her purse at me. He ran out, leaving me with his wife. I was frozen in place, sitting on the bench topless and covered in stinky stickiness. I tried to speak, but no sound came out. She screamed at me that I was a whore and to get the fuck away from her husband. I tried to tell her. She told me she was going to make my life miserable in Seattle. I just looked at her and said, "It already is miserable." I wiped myself off on the dress I was about to try on and left. Numb. Shocked. Dirty. When I got home, I washed and washed myself, certain my father would see what had happened and blame me.

Up until recently I considered that my first sexual encounter. It took me until a few years ago to realize that wasn't sex, it was child molestation. And anyone who has experienced this, I'm truly sorry. You did not deserve it. You did not ask for it. You were not at fault.

I'd pushed away the shame feeling for years. Whenever I thought of it, I could hear my father telling me that I shouldn't wear nail polish because God would see the true dirt under my nails. That's what it made me feel like. Dirt under fingernails. Right now, by writing this, I am looking at the dirt. I am unpacking it.

For a long time, I thought that you had to be penetrated by a penis for it to be rape. I was wrong. Anything that isn't sexually consensual is sexual assault. Fingers, mouths, penises, if they are in you or on you without your consent, that is sexual assault. Once I was at a gay club, standing on a chair to watch a performance, and a gay man stuck his finger inside of me and he said, "Oh, I've always wanted to see what that was like." Only more recently have I realized that qualifies as digital rape. It was violation and assault all under the assumption of the guy can take it because you drove him to it by causing him to *want*.

What that poor young girl experienced on that set was rape, too, as far as I'm concerned. She told me about it at a cocktail party. I could tell it caused her deep pain and it broke my heart.

When you're a girl or woman, you're told that your greatest reward is being recognized for your beauty, and being desired. You're supposed to play it up every time you're in the public eye, but then downplay it offscreen, in your private life, because you don't want to make anyone uncomfortable and you don't want to attract the wrong kind of attention. That poor girl's been brainwashed into thinking her beauty is all she's worth, because from the time she was a little child she's been getting compliments about her beauty. And it becomes something you think you owe other people.

To all the beautiful girls out there, and the beautiful boys: you don't owe anybody anything because of how you look. Society gives men a free pass: "They can't help themselves." "You drove them to it with your beauty." I was told that so many times: "They just can't keep their hands to themselves . . . Oh, that poor man can't handle it." Well, fuck you, that's called assault. Nobody calls it what it is in this society, but it's time to start.

We are not disposable, we are not "just girls." The women of Juarez, the missing indigenous women in Canada, the "child brides" kidnapped by Boko Haram in Nigeria: all just girls. It's time to stop thinking of us as just girls. We're not "just anything." We are fully fledged human beings. Consider our lives.

WE ARE BRAVE

At this point you might be thinking to yourself, *What is this woman complaining about? Doesn't she know how lucky she is?*

I heard this throughout my entire career: "You are so lucky."

Those four words built invisible walls around me that kept me isolated, scared, resentful, and lonely for the better part of twenty years.

I consider my luck to have not contracted leprosy or being a blind homeless child in Calcutta. Fame to me isn't luck. Yes, there are people who crave being idolized, recognized, celebrated for nothing else more than existing. That was not and is not me. For me, fame was a corrosive force, something to be survived. There is so much more that I haven't written here. Everyone deserves dignity, everybody.

So while I am grateful for the experiences I've had—set camaraderie, travel, adventure—I don't believe I owe Hollywood a thank-you letter. I owe it a kick in the proverbial head. Thanks for not protecting me, thanks for trading me, thanks for hurting me. But, sure, go ahead, call me lucky.

To the writers who create how we see one another and ourselves: You're responsible. What you write forms the thoughts and self-images of billions of humans. Take care with your words and your images. Grow up. Get smarter. Think deeper.

I have repeatedly asked the head of a writers union—a very, very intelligent man, and very liberal—if I could speak to his member body. I wanted to speak to his writers, men AND women, peer to peer. I wanted to talk to these writers about their consistent misrepresentation, most especially of women. And every single time I asked the head of this union about speaking to his writers, he said, "I'll have to ask my women's committee." This was about writing better characters, not just for women, for everybody. But this is typical of a male liberal who is unwittingly sexist and why we are where we are. I'll have to ask my women's committee. Fuck off with that. How are you supposed to fix the messaging system if no one can speak to it?

Then there is the representative problem. The managers and agents. Based on everything I've experienced and seen, far too many don't have the bandwidth to remotely understand the nature of an artist. If they did, they would protect and nurture artists, and then they would actually be of value to us. Too many agents seem to think they are the stars, that they hold the power. It's delusional. I think anybody who profits off another human body is nothing more than a pimp. I had men negotiating how long my breasts and ass could be shown on-screen. On the street that is a pimp. In Hollywood, they make millions in commissions. That is a form of human trafficking.

And the Screen Actors Guild (the actors' union): You are a part of this. Don't you get it? You should be saying: "We're a union. We're going to protect our members on set, especially the most vulnerable, the females and children. We're going to make it mandatory that there's equal pay for equal work." You are missing out on an opportunity to set standards and right things. I paid you an incredible amount in fees so that you would protect my rights. But you didn't. An actress is an employee just as she would be in any other job in any other industry. There should be a hotline for anonymous tips of abuse on the set and abuse of power. Our union would really make positive change by doing so. It is inexcusable how you let these crimes happen to your member body.

And directors: I have already said much about you guys in these pages, your infantile tantrums, the boys' club entitlement. Please learn once and for all that a girl, a woman, is not just this thing in a skirt to get other dudes to come to your movie. Aren't you tired of being a cliché? We are so tired of you. And STOP using rape as a tool for your storytelling; it is damaging and causes trauma.

Most of you guys are making mediocre movies at best; at worst, you're making shitty fucking schlock. Very few of you are making classics because you're not good enough. Because you don't have the skills as a human being. A great director must be great at empathy and multitasking, classically female strengths. Maybe it's why so many of these men suck.

One time I was in the back of a car, and Quentin Tarantino and Robert Rodriguez were in the front seats. I wanted to see the extent of the male director's ego. I thought I would just do my own little test. I decided to see how slimy I could be, to see if these guys would buy it. I figured no way, since I'm not exactly known for being saccharine. So I laid it on thick, going on about "what it's like for me as a woman to be 'allowed' to be in this car with you two men that are like walking gods. Everyone worships you, wow, what's it like to be you, the strongest, toughest guys on earth with the hardest jobs blah blah blah." I was so over the top with my delivery, thinking, *Come on, guys.* Come on. *At any moment you're going to realize and laugh and tell me to shut the fuck up.* Nope. Their shoulders puffed up, and their plumes came out. It was as if I were being choked by feathers made of ego. I hated them.

They took it 100 percent seriously because they are surrounded by people (and women producers) who coddle them and kiss their ass.

Women in the industry: Ladies, you need to step up and realize the men are never going to scoot aside for you; they're not going to offer you a seat at their table, so build your own damn table. Figure it out, you can do it. In all other respects in your life, you're a boss, right? So figure this one out. You got this.

The female producers who are hired to coddle the poor

male directors through the hard, hard life of directing just pacify them at all costs. There are so few women directors who work, but there are a lot of women in producing because, of course, legions of these baby man directors want someone to mommy them. It's such a cliché.

The women who can invest in women writers and directors: Why are you still investing in men? Why do you keep playing by their rules? Why do you keep supporting a system that's not supporting you and your kind? To women producers: Why are you not hiring women en masse? Why are you female agents not representing as many women as you can? Do it on purpose, do it every time. Be brave. If one doesn't work out, try another. It's not like many men haven't worked out.

Actresses, female actors: You are not a commodity; you are artists. Go to the female producer if there is one and demand that she stand up to the director on your behalf. If there's no female producer, you have to find somebody who will stand up for you and become your ally. Speak out about the torment and abuse in auditions and on sets. Lodge complaints against your union if they aren't protecting you and others. Walk out of auditions where you're lined up with girls all wearing bikinis and given numerical ratings by douche-bag directors. Say no to demeaning, objectified, cliché roles, and say *why* and MAKE IT PUBLIC. Say no to being a show pony on the red carpet. Say no to rape scenes. Demand respect and equal pay.

To men in general: I think it's high time that you take a long look in the mirror.

Your kind are the number one danger to women and children. You are the number one danger to animals and the planet. We're facing extinction because of you. You are the number one reason why we have wars, mass deaths, rape, molestation, torture, and a host of other ills. In other words, the problem is your kind, maybe even you. Know that you and your kind must change. It is your responsibility to do the work. No one else is to blame, it is you who **must** break the cycle. Just like how racism is not POC's problem, misogyny is not women's job to fix, it is yours.

Look at ways in which you're privileged. Even with my most "awake" male friends I see casual entitlement, without their being aware of it in any way. From the first breath you draw as a male child you're stamped by society as superior. You can be in the most down-and-out, fucked-up situation, but you're still better than a girl in that same situation. Do you ever wonder why that is? Why? I just can't do the math. I see no evidence of superiority. I don't think having a penis makes you superior; in fact, I think it's what makes you vulnerable and your vulnerability scares you. The fear that men push down causes all too many to lash out to prove their toughness. It is not working. Change.

Don't get me wrong, I feel bad for the box men get put in, too. Men are born into such a narrow idea of what they are supposed to be. Snore. I can see this in my own life, with my father. He was raised on this steady diet of John Wayne films by his macho Korean War–fighting dad. His dad told him that a real man doesn't cry, a real man doesn't complain, a real man, a real man, a real man . . . ad infinitum. "Go watch John Wayne, you'll never see him be weak. Ever." How many men grew up with this false idea of strength? For years Hollywood has been responsible for showing men how to be this bullshit ideated version of themselves. It's time to knock that one down.

Aren't you bored with being the tough loner, the jock, the nerd, the womanizer? Wouldn't you like to develop into a deeper, more nuanced human? You're living an illusion that's generated by a mass propaganda machine. Don't you want to see the Matrix for what it is and get out? For those of you who are out, go further.

I can't wait for us all to be just humans. No gender. No stereotypes. Just humans. But we need to unpack the traditional indoctrinated thought first.

Ahh, the patriarchy. The patriarchy to me is the equivalent of male society resting on a La-Z-Boy recliner, basking in its comfort, but don't you realize it's a trap? Why does a woman who's strong make you less? It's not a zero-sum game. My equality and respect doesn't leave less respect for you: it's not fucking pie. There's enough creativity, love, resources for all of us, if you'd just stop hoarding stuff. You're safe, okay? No one is trying to get you, it's you who are trying to get us. Look at the statistics. What if you could no longer be ruled by insecurity and fear? What if you too could be free?

To women: Why are you still apologizing? Why are you constantly saying sorry, for just existing? Occupy space. Take it. Aren't you tired of not knowing your worth? Aren't you tired of having your value equated with your fuckability and your looks? Aren't you tired of competing with other women? Aren't you tired of societal constructs benefiting males that encourage us to turn against each other? Why not turn and help the woman or the girl next to you? Instead of hiring the guy, hire the girl. Do it on purpose. And if it doesn't work out the first time, try again. Let's do our own version of Affirmative Action. It's not sexist, it's right and equitable. And long, long, long overdue. Do not play by their invented rules. Stop.

The reason I know my own worth now is because I see all that worth in you. I see how much you are worth and how much you don't see it yourself. If I can do it, anyone can.

If you're called a man hater, bitter, ugly, not hot, a misandrist, whatever other label they can think of, would you not survive? Yes, you would. I have been called every name in the book and I have survived.

I'm called bitter when I fight for equality. I don't know what equality in the eyes of the law and equal consideration has to do with bitterness. That is just another label used to shut me and those like me up. But who cares if slow-witted men label us as bitter? They're always going to label us something; we may as well have a cause.

Men and women: You can both be feminists. It just means you're for equal rights and equal pay. If you believe women should have equal rights and equal pay, congratulations, you are now a feminist. Pretty simple, really. When you are pro-equality, you get your humanity back. When you get your humanity back, you start realizing that you are in fact in control of the narrative. It's a place of power. Feminism has for years turned into a dirty word, but it's not. The men in power and in media made it a dirty word to serve *their* purposes, not ours. Look at who's benefiting from you not recognizing that you are a feminist, because it's not you. So be one. There is no price for admission, just a free mind and a desire for equality. Go, you.

And to those who are Survivors: You need to know that it is YOU who is awesomely strong. Look at what you've done and how far you have come. Go, fucking you. Badass. Look at how you've gone forward in your life despite every goddamn message you get every goddamn day that you are not good enough. Because you know what? You are. And I love you and honor you.

When I was in the Cult of Hollywood, I'd put myself down and say I had useless talents. Useless talent number 72, useless talent 75, et cetera . . . I was a strong writer, but it wasn't my official job, so to my mind, it was just a useless talent. I am a damned good photographer, but never pursued it as an additional career because I already had one. Photography is what others did professionally, not me; I was already an actress. And then there was useless talent 47: music. I love singing, and I'm a strong lyricist. But no one told me I could do whatever I wanted to do. I never got that message, so I gave it to myself, and now I give it to you. Be you. Be it loud and proud.

Do whatever you want as much as you can. Do whatever you want to do, because this is your life. Live a life of adventure, tiny and big. Do weird shit. There is no reason you have to live your life to the rhythm of others. If you feel like staying up at night and sleeping during the day, figure out a way to support that and do it. This is your time, this is all you get, as far as we know, so why spend it making others happy if it's not making you happy?

If you were your true self, what would you be? Wouldn't it be amazing if we were all our best selves? I think that's a noble thing to think about and work toward. We can be better by thinking differently. Whatever is different about you is what makes you amazing. Others will try to homogenize you for their own comfort level, because God forbid discomfort. Fuck that. Do not bend yourself to make others feel taller. When someone comes at you telling you what you are and slapping a label on you, simply ask them why. Why would you say that? Make *them* think.

Be creative at whatever you do and with whatever you are. It does require bravery, but I believe in you; I know you've got it in you to be better. I know you've got it in you to be brave.

P.S.

This is where I give you the postscript of my life, where I'm at now. I'm at a pretty great place. I have enjoyed becoming braver and braver. Being more comfortable in my skin. I have enjoyed using all my artistic talents that were once used as a muse and an inspiration for others. I shoot commercial campaigns. I'm directing films. I'm doing an album. I'm launching an incredible labor of love, a skincare line that challenges what the beauty industry tells you you need. It's called The Only Skincare. I've been working on this special three-in-one formula with my aunt Rory for the last eight years. It's an amazing line, completely revolutionary in its simplicity and efficacy. It makes me so happy that I'll have another way of doing things differently. The beauty industry also needs to change its messaging system, and by creating something truthful that really works, I'll be embarking on that journey soon. Also, as I previously mentioned, my album, *Planet 9*, will be released around the same time as this book. The book and album really do go together, and both are about freedom of thought.

Everyone thinks, unless they see you on a screen, you must just cease to exist.

Well, I'm here to tell you that's not true. I live an amazing, really varied life filled with travel and work and joy.

I spent most of my life, most of my adult life, working in service of others, but being silenced by doing so. I no longer do that. I can't. I have to use my platform for good. The work now is work that I love and work that I'm free to do, and work that, frankly, I excel at. It took years for me to master, but I did. I want to spark conversation and in that way effect change.

For a long time, I thought I got my strength from my dad, but I was wrong. I got it from every woman who came before me in my family. I got it from my brilliant, super-political mom. My aunt Rory, my aunt Kelli, my aunt Debby, my grandma Nora. My great-grandmother Ruby, who was one of the first women to go to the University of New Mexico. She had men throwing things at her every day on the way to school, but she persevered. The women in my family, they just get on with it, as so many of us do. Here's to all of us living in a new world, one that's different, a world where we don't have to just get on with it, but a world where we are free. Here's to us all demanding more.

Being brave doesn't mean you are not scared, it just means you do the scary thing anyway. So look at what you consume and choose wisely. Email, tweet, demand different from those who are pushing a damaging narrative. And if they don't change, refuse to buy what they are selling. Push back at the system where you see injustice happening. It is more important than ever that we grow and grow fast. Our lives are at stake, our minds are at stake. See how women are treated in life and stand up for us. Don't join in with the degradation too often heaped on us, and stop those doing the degrading. Don't play the machine's game. Be better. Think different. I know you can. I know you can change the world, starting with yours, just by being brave.

ACKNOWLEDGMENTS

Thank you to all those who have shown me kindness and consideration along the way. You know who you are.